PROBLEMS OF FORM

WRITING SCIENCE

EDITORS Timothy Lenoir and Hans Ulrich Gumbrecht

Contributors

Dirk Baecker

Giancarlo Corsi

Elena Esposito

Michael Hutter

Klaus P. Japp

Niklas Luhmann

David Roberts

Karl Eberhard Schorr

Fritz B. Simon

Rudolf Stichweh

Helmut Willke

PROBLEMS OF FORM

Edited by
DIRK BAECKER

Translated by
MICHAEL IRMSCHER, *with* LEAH EDWARDS

STANFORD UNIVERSITY PRESS
STANFORD, CALIFORNIA
1999

Stanford University Press
Stanford, California

© 1999 by the Board of Trustees of the
Leland Stanford Junior University

Printed in the United States of America

CIP data appear at the end of the book

CONTENTS

Contributors   ix

Introduction   1
DIRK BAECKER

The Paradox of Form   15
NIKLAS LUHMANN

Self-Reference in Literature   27
DAVID ROBERTS

Sign as Form   46
NIKLAS LUHMANN

On the Analysis and Use of Form in Logic   64
KARL EBERHARD SCHORR

Two-Sided Forms in Language   78
ELENA ESPOSITO

The Form Game   99
DIRK BAECKER

The Early Form of Money   107
MICHAEL HUTTER

The Form of the University   121
RUDOLF STICHWEH

The Contingency and Necessity of the State   142
HELMUT WILLKE

The Form of Protest in the New Social Movements   155
KLAUS P. JAPP

The Dark Side of a Career   171
GIANCARLO CORSI

The Other Side of Illness   180
FRITZ B. SIMON

Notes   201

CONTRIBUTORS

DIRK BAECKER is professor of organizational theory at the University of Witten/Herdecke, Germany. His work includes *Information und Risiko in der Marktwirtschaft* (1988), *Die Form des Unternehmens* (1993), *Postheroisches Management: Ein Vademecum* (1994), and *Poker im Osten: Probleme der Transformations-gesellschaft* (1998), as well as "The Writing of Accounting" (*Stanford Literature Review* 9, 1992).

GIANCARLO CORSI is a researcher at the Centro di Studi sul Rischio at the University of Lecce, Italy. He is the author of *Sistemi che apprendono: Studio sull'idea di riforma nel sistema dell'educazione* (1997).

ELENA ESPOSITO is a researcher in sociology at the University of Urbino, Italy. She is the author of *L'operazione di osservazione: Costruttivismo e teoria dei sistemi sociali* (1992), as well as "From Self-Reference to Autology: How to Operationalize a Circular Approach" (*Social Science Information* 35, 1996) and "Observing Objects and Programming Objects" (*Systems Research* 13, 1996).

MICHAEL HUTTER is professor of economic theory at the University of Witten/Herdecke, Germany. He is the author of *Die Produktion von Recht: Eine selbstreferentielle Theorie der Wirtschaft, angewandt auf den Fall des Arzneimittelpatentrechts* (1989), as well as "The Value of Play" (in Arjo Klamer, ed., *The Value of Culture*, 1996), "The Impact of Cultural Economics on Economic Theory" (*Journal of Cultural Economics* 20, 1997), and "On the Consumption of Signs" (in Marina Bianchi, ed., *The Active Consumer*, 1998).

KLAUS P. JAPP is professor of sociology at the University of Bielefeld, Germany. He is the author of *Soziologische Risikotheorie: Funktionale Differenzierung, Politisierung und Reflexion* (1996).

NIKLAS LUHMANN (d. 1998) was professor of sociology at the University of Bielefeld, Germany. His work includes *The Differentiation of Society* (1982), *Love as Passion* (1986), *Ecological Communication* (1989), *Essays of Self-Reference* (1990), *Risk: A Sociological Theory* (1993), *Social Systems* (1995), and *Die Gesellschaft der Gesellschaft* (1997).

DAVID ROBERTS is professor of German at Monash University, Clayton, Australia. He is the author of, among other works, *Art and Enlightenment: Aesthetic Theory After Adorno* (1991).

KARL EBERHARD SCHORR (d. 1995) was professor of pedagogics at the University of Hamburg, Germany. Together with Niklas Luhmann he edited a series of volumes on the sociological systems theory of education: *Zwischen Technologie und Selbstreferenz* (1982), *Zwischen Intransparenz und Verstehen* (1986), *Zwischen Anfang und Ende* (1990), *Zwischen Absicht und Person* (1992), and *Zwischen System und Umwelt* (1996).

FRITZ B. SIMON is a family therapist and director of the Heidelberger Institut für systemische Forschung, Therapie und Beratung. His work includes *The Language of Family Therapy: A Systemic Vocabulary and Sourcebook* (with Helm Stierlin and Lynn Wynne, 1985) and *My Psychosis, My Bicycle, and I: The Self-Organization of Madness* (1996).

RUDOLF STICHWEH is professor of sociology at the University of Bielefeld, Germany. He is the author of *Zur Entstehung des modernen Systems wissenschaftlicher Disziplinen: Physik in Deutschland 1740–1890* (1984) and *Der frühmoderne Staat und die europäische Universität* (1991), as well as "Science in the System of World Society" (*Social Science Information* 35, 1996) and "Professions in Modern Society" (*International Review of Sociology* 7, 1997).

HELMUT WILLKE is professor of sociology at the University of Bielefeld, Germany. He is the author of, among other works, *Systemtheorie I–III* (1982–95), *Ironie des Staates: Grundlinien einer Staatstheorie polyzentrischer Gesellschaft* (1992), *Benevolent Conspiracies: The Role of Enabling Technologies in Reframing the Welfare of Nations* (with Carsten Krück and Chris Thorn, 1995), and *Supervision des Staates* (1997).

PROBLEMS OF FORM

DIRK BAECKER | **INTRODUCTION**

Mathematics has often attracted sociologists. They use mathematical methods to work on statistical data. They rely on the power of mathematical reasoning when they distinguish variables and constants. They learn from systems of mathematical equations how to treat the different (the two sides of the equation) as identical (with respect to the equal sign). They refer to mathematical equations when they claim that variation may be a more important concept than causality. And they take mathematics as a model of how the most abstract thinking nevertheless is able to refer to the most concrete examples. Mathematics is the model of simultaneous closure and openness.

Though sociologists are intrigued by mathematics, they are never sure whether mathematics itself in all these characteristics just mentioned is not simply an improbable product of the same characteristics already provided by language and writing, that is, by communication. One may indeed study certain characteristics of communication by studying mathematics, yet it is communication that is the answer to the question of how these characteristics come about, and not mathematics. There is no way to render sociological thinking exogenous by asserting the existence of some external reason of reason, incorporated in mathematics. All mathematical achievements are devices that bear witness to the peculiarities of communication.

Social-systems theory is fascinated by a mathematical calculus that seems to say a lot about this hitherto rather hidden familiarity between mathematics and communication.[1] George Spencer-Brown, a British professor of mathematics, wondered while developing a counting machine for British Railways why formal logic could not handle the engineer's use of imaginary values in calculations. In response, he developed a calculus that refounded logic on the basis of mathematics. His works also contain glosses, prose stories, and poems that he published under the nom de plume James Keys. Spencer-Brown's calculus of indications, presented in his book *Laws of Form*,[2] exhibits all the strengths of mathematical computation, yet at the same time captures communication's capacity for, and anchoring in, ambivalence and ambiguity. Spencer-Brown's is a mathematics of self-reference and of distinction, which for the first time in mathematics explicitly includes the observer in the operations the calculus performs.[3] That, of course, was reason enough to welcome his calculus at once into second-order cybernetics and to claim that Spencer-Brown had delivered the mathematics for understanding self-referential ("autopoietical") systems.[4]

Indeed, Spencer-Brown's calculus presents us early on with a communication, preceded only by a definition and the two axioms of the calculus. The communication reads, "Draw a distinction."[5] Only by drawing a distinction is one able to indicate something and thus know what to refer to. If a reader follows the calculus through to the end (which leads back to the beginning, which now looks different, since you not only can do what you do but also can watch what you do), he learns that he has participated in the construction of a form that lets him choose whether he wants to follow, or not, this communication or another one. At the end he discovers how the distinction he drew at the beginning is to be understood. And he discovers the two ways to handle distinctions. That is why the definition, which captures the understanding of the distinction, and the two axioms are part of the calculus itself and are presented as its first chapter under the title "The Form."

The definition reads, "Distinction is perfect continence."[6] A distinction contains everything: the indication that the distinction makes; the nonindicated rest of the world, which the indicated is distinguished from; and the distinction itself, separating the states indicated from the states nonindicated. It even contains, as Spencer-Brown goes on to elaborate, the motive, the content, the value, and the name of the distinction. This means that one cannot proceed to construct anything, indeed to do anything, without drawing a distinction. And it means that no presuppositions are necessary for drawing a distinction except, as you discover at the end, the fact of always already being involved in the endless process of drawing distinctions. Thus the distinction is not identical to the indication it makes but contains it, as it contains the shaded rest of the world, including the distinction itself, drawn by the observer who follows, or rather enacts, the communication. Everything the distinction contains is called its "form."

Obviously, there are two and only two (but do not take this proposition literally) ways to handle distinctions. One may either "call" a distinction again and thereby confirm its value, or one may decide not to follow its indication and look instead at its nonindicated states, thereby "crossing" the distinction. Thus axiom 1 reads, "The value of a call made again is the value of the call." And axiom 2 reads, "The value of a crossing made again is not the value of the crossing."[7] Both axioms elucidate that the distinction is understood as an operation, not as a state or a thing. It is doing, performing, or creating the form, not representing, classifying, or symbolizing it. We are dealing with an operational calculus, not with a propositional calculus. Spencer-Brown has read Wittgenstein even more than he has read Bertrand Russell. A distinction is an operation. That is why it may be "called" or "crossed." By calling it, you confirm it. By crossing it, you cancel it. By confirming it, you end up with the same indication, save that it is now a repeated indication, which indeed may change something (but not inside Spencer-Brown's mathematical calculus). By canceling it you end up with empty hands, since now you have chosen the unmarked state of the world, or the nonindicated state of the distinction, instead of the indication. That, by the way, is why Spencer-Brown's calculus shows some affinity with Buddhist meditational practices. These practices consist in teaching the possibility not only of following distinctions but also of abolishing them, thus turning them from necessary into contingent ones.

By not only performing distinctions but teaching something about them, Spencer-Brown's calculus already exhibits a third way to handle distinctions. It both uses and talks about distinctions, for distinctions can be "re-entered" into the realm they distinguish. Hence an observer can not only choose between two sides but observe the "form" that makes such a choice possible and observe the observer who hides behind the distinction and who called out "Draw a distinction" in the first place.[8]

Thus Spencer-Brown's calculus shows how one-sided—that is, necessarily asymmetrical—indications (or biases) are introduced by two-sided distinctions, which are based on a three-valued form: The first value is the marked state of the indication, the second the unmarked state of the excluded rest of the world, and the third the observer drawing the distinction and separating two states of the world that may, for another observer, be indistinguishable. Operations are necessarily asymmetrical because they consist in preferring a certain (and chosen) way to proceed over ways not to proceed. But they can return to symmetry when one observes their chosen form. Observation of that form turns them from "crosses" into "markers."[9] As markers, distinctions represent a boundary, but they no longer draw it.

This sketch of some features of Spencer-Brown's calculus makes apparent that social-systems theory did not necessarily become interested in it for its mathematical or logical contributions. There are two reasons for systems theory's interest. First is the explicit inclusion of the observer in the operations he performs, together with the possibility of observing, by using the indication (i.e., distinction) of the "form" of his distinctions, how he does what he does and which possibilities he leaves open for choice (for himself as well as for others). Second is that both the concatenation of the operations performed by drawing distinctions and observations of these operations performed by drawing other distinctions constitute a perfect example of communication. Any observer drawing a distinction necessarily leaves open the three options for handling the distinction: to call it again, thereby accepting its motive, value, and content; to cross it, looking for different motives, values, and contents; or to re-enter it, perhaps in order to examine more closely its motive, value, and content. Drawing distinctions and thereby creating the space of choice is exactly what communication does,[10] restricting the realm of the possible yet exposing the very act of restriction, together with the realm of the nonrestricted (yet unspecified) and the observer responsible for the restriction (and its exposure).

The essays in this volume make selective use of Spencer-Brown's calculus. Their main purpose is to inquire into methodological and analytical possibilities of taking seriously "the facticity of the observer," in Niklas Luhmann's phrase. This means at least three things. First, the social realm is understood to be a realm of communications performing observations. Observations are not something done after the fact or after the deed. Rather, they are the fact and the deed itself. Any event, any action, any communication is taken to be an observation concerning a certain state of a social system, which changes the state and thereby reproduces the system. Social phenomena draw distinctions between possible states of a system and thereby in-form the system. In this way social systems are analyzed as "cognitive" systems, if "cognition" means drawing a distinction between self and non-sense.[11]

Second, these observations made via distinctions, if operative inside social systems, are themselves observed. Social systems emerge on the level of second-order observations. This means that social phenomena may be analyzed as phenomena handling in different ways the options presented by the observation of the form of a distinction. Social systems must themselves decide: whether, how, and by whom they demand the mere calling again of their distinctions; whether, how, and by whom they tolerate the crossing of their distinctions; and whether, how, and by whom they support the observation of the form of their distinctions. The fact that the laws of form have only now been discovered says something about how social systems condition themselves by enforcing and restricting their use of their distinctions.[12]

Third, sociological inquiry is itself carried out by distinctions that include what they indicate and exclude what they don't. Ever since Auguste Comte advised becoming a "positive" scientist (rather than a priest referring to gods, or a philosopher referring to abstract ideas),[13] sociology has been explicit about the distinctions it uses to observe society. And ever since Karl Marx proposed doing a "critical" analysis of society in terms of distinctions the society uses "ideologically," which may fall prey to a "revolution" that uses different ones (but Marx cautiously is not too explicit about these new distinctions), sociology has aptly singled out the distinctions in which a society entangles itself. Yet somehow these two competencies have never quite meshed. Sociology has always oscillated between programming its own distinctions and criticizing society's distinctions. It found only a temporary rest when it adopted indications like "community," "authority," "status," "the sacred," and "alienation."[14] These indications were of use both in making explicit sociology's distinctions and in criticizing certain states of the society, be they the states indicated or the corresponding states nonindicated ("society," "the self," "achievement" or "contract," "the secular," "emancipation").

In adopting the calculus of Spencer-Brown as a methodological and analytical device, sociology has placed itself in a paradox. On the one hand, sociology is forced to accept a ban on self-exemption. Everything it says about an observer acting inside society must be valid for sociology as well, sociology being nothing but another observer acting inside society. On the other hand, sociology becomes an observer using distinctions to observe social systems using their distinctions to steer communications that are themselves distinctions. The epistemological conundrum of this situation is that you can never know how your distinctions are able to capture the distinctions drawn in the field, since they are precisely your distinctions. You can only take comfort in the fact that everybody else is doing the same. That is why the dynamic stability of social systems can be gained only on the level of second-order observations: observers observing how observers deal with objects that are nothing but references to observations stabilized by observations.[15] A decisive step toward acknowledging this type of sociological third-order observation of social second-order observations is the use of second-order concepts now widespread in sociology.[16] Yet methodological awareness of these concepts is still in its infancy.

The essays in this volume make different uses of the calculus. In "The Paradox of Form," Niklas Luhmann presents his way of reading the "laws of form." One of his interests is to watch how Spencer-Brown develops a notion of form that is separate from such venerable opposites as "matter," "substance," or "content" and that works self-referentially on itself, containing its two sides and the distinction separating them. Some critiques maintain that this proves systems theory is moving toward universal abstraction and doing away with matter altogether.[17] This critique understates how "matter," "substance," and "content" move into the form of the distinction, which calls out for a very material and empirical analysis of the operation of distinction carried out by observers and manifest in the possibility that they will be observed themselves. Such an analysis reveals that an all-encompassing form is necessarily paradoxical since it cannot "contain" everything without being different from it, which obviously it cannot be.[18] Luhmann therefore calls form itself a paradox, being, as it were, the unity of a distinction. Like any paradox, it calls out for a distinction (that is, for "itself") to take the place of the paradox, make it invisible—and be "deconstructed" in order to rediscover the paradox.[19]

Problems of form ensue from the fact, typical of modern society, that distinctions not only are used to produce and reproduce certain types of communication but are observed, at the same time and by different observers (who may be users themselves sometime later), with respect to their form—that is, with respect to their paradox, contingency, and continence (i.e., self-control). The problem is that these distinctions must be used nonetheless. Only other distinctions, for which the same caveat would be in order, could substitute for them. David Roberts, in "Self-Reference in Literature," analyzes romantic irony as an instance where this problem began to occupy the center of reflection. Yet it was rendered invisible by being transformed into the question of how the unmarked states of a distinction are to be presented via distinctions and into the observation that the unsolvability of this question was to be understood as the source of poetic imagination, which thus could never dry up.

Problems of form usually are disguised by either communicative bans or semantic blurring. Traditional society preferred the ban, and then the ban was indeed sufficient, enabling first-class semantic performances such as mystical knowledge, theology, or poetry to handle paradoxes. Modern society seems to prefer semantic blurring, which has the advantage that different observers can entertain both belief in an ability to put things clearly, simply, and straightforwardly and the knowledge of the paradoxical foundation of our knowledge of the world. The notion of "sign" is an example of semantic blurring. Niklas Luhmann, in "Sign as Form," analyzes the confusion between, on the one hand, the sign taken to be the signifier and, on the other, the sign considered as the unity of the distinction between the signifier and the signified. If you take the sign to be the signifier, you end up with the "postmodern" belief in the loss of the referent because signs are "arbitrary." If you take the sign to be the unity of the difference between signifier and signified, you may notice that, though indeed the sign is without a referent, it is at the same time the device (Peirce's "interpretant" or "third") that establishes a relation between signifier and signified, conceived as the establishment of a relationship of reference. Of course, the referent is motivated not by the world it refers to but by the play of signs that provide references to the world.

In thinking about problems of form, one encounters predecessors and relatives. Here C. S. Peirce's "interpretant" is as important as Ferdinand de Saussure's arbitrariness of the sign; Gregory Bateson's "difference that makes a difference" as important as Jacques Derrida's *différance*. All these concepts ask how recursive iterations are possible in operations that presuppose a whole world yet gain everything they need to reproduce themselves (that is, their unity and their distinction) from their own reproduction. The answer to this question is not the ancient idea of a creation ex nihilo but the new idea of autopoiesis.[20] Autopoiesis is a kind of production that combines available ("internal") and unavailable ("external") factors of production. Such a notion requires a reconsideration of many semantic achievements in the self-explanation of society—requires that they be deconstructed and reconstructed, the deconstruction revealing the paradox, the reconstruction revealing its operational translation into recursive distinctions.

As an example of such a shift from symbolically to formally analyzing semantics,[21] in "The Analysis and Use of Form in Logic" Karl Eberhard Schorr recalls Gottlob Frege's notion of writing as both the procedure and the assertion of modern formal logic. He shows that since Kant logic has been less interested in assuring faultless deduction and monitoring self-reference lest it become paradoxical than in establishing and exploiting an order that contains sufficient redundancy to offer both the problem that there be a next step and its solution. The emergence of undecidabilities, then, both reveals a lot about the structure in which one moves and requires an "inconsequential" recourse to "unavailable" reference points, which thereby become necessary parts of the game.[22] Kurt Gödel's discovery is one version of this, Alan Turing's problem of halting another.

A similar problem is taken up by Elena Esposito in her essay, "Two-Sided Forms in Language." She starts by showing that the basic operation of distinction in Spencer-Brown's calculus is itself not a simple but a complex operation: an operation that supports itself in (paradoxical) self-distinction from everything else.[23] She maintains that a comparably complex operation is the foundation of language. Language must provide the means to differentiate itself from other types of behavior and from other ways of referring to the world and must gain and maintain the ability to connect actual language events to previous and possible further ones. These means are assured by two distinctions, which co-evolve and are not reducible to one another. The distinction between self-reference and external reference provides for the differentiation and autonomy of language; the distinction between indication and distinction (in Spencer-Brown's sense) concerns the interconnectability of basic language events. Language itself assures the working of these two distinctions because it gains both isolation and redundancy by being "doubly articulated," so that words and sentences are decoupled from what they refer to, yet are recoupled to each other to organize the references they are expected to maintain.

My essay "The Form Game" takes up Bateson's notion of play. It demonstrates how Spencer-Brown's figure of the re-entry of the distinction into the realm it distinguishes may serve to show that metacommunication is not an additional achievement of communication that sometimes is called up and sometimes not, but the very foundation of the possibility of communication. A child learns to communicate by learning to distinguish between "cross" and "marker," that is, between distinctions drawn and distinctions exposed. Thus playing in Bateson's sense, which introduces the frame of communication into the communication, is the initiating mechanism for communication and is perhaps the most elementary social practice. Different social forms may be distinguished by watching how the play of communication is entertained and restricted.

These rather general chapters are followed by others that look at more specialized social forms. Michael Hutter, in "The Early Form of Money," describes how early Mediterranean coins came about: lumps of metal already in use were first marked and then sealed (closed) by having their backsides marked as well. This meant that these forms are not to be reopened, that is, changed in value by having metal removed or added. Two questions ensue. First, who will respect the marks? And second, who makes the marks that are to be respected? Both questions were answered by a peculiar game of revealing and concealing the observer who provided for the marks, these marks sometimes being "private" ones that related their validity to some "public" figure of respect, sometimes "public" ones, legitimated by being accepted as legal tender in some clearly demarcated "private" (royal) household. A certain blurring of the distinction between the private and the public, which nevertheless remains a distinction of vital importance, may be responsible for the prevalence of this early form of the coin, which can still be found today.

A reconstruction of blurred distinctions may be important to understanding the university as well. Rudolf Stichweh, in "The Form of the University," looks at the astonishing multiplicity of educational institutions that call themselves "universities" and at the equally astonishing continuity of the name "university" and of several universities that are among the oldest organizations of the society, third only to the state and the church. Stichweh examines distinctions that maintain the university: on the one hand, external distinctions between the university and something else (most notably state and church) and between universalism and particularism, and on the other hand, internal distinctions. Examples of the latter are Humboldt's distinction between *Forschung* (research) and *Lehre* (teaching) and, much more important today, the distinction between what, in the university, can be reformed and what cannot, a distinction used both to invite state reform and to keep such intrusion at bay. These operative distinctions conceal the distinction that defines the societal function of the university: the distinction between taking care of the young and letting them go.

All social problems of form (are there any others?) have to do with distinctions originally thought to be necessary, yet revealed to be contingent. The state, as Helmut Willke discusses in "The Contingency and Necessity of the State," is a famous example of how the laws of form work. Hegel took up the state as the institution of last resort in a civil society, which viewed the person as a mixture of nature's necessity and nature's arbitrariness (i.e., freedom). To claim the necessity of the state as the unity of all difference is a contingent solution to the problem of contingency in general. But only at the end of the nineteenth century was the contingency of politics acknowledged and the state accepted as a contingent means to control this contingency. Following Rorty's notion of "irony," Willke suggests that our way of observing (in its double sense) the state can only be ironical, a way that accepts contingent solutions to societal problems as contingently necessary, solutions that never are meant to be final but that keep the space of further possible action open, or even better, enlarge it.

Problems of form do not go unnoticed in society. That is why they are called problems of form. They result from observation. Yet that does not mean that all observation of form necessarily is self-reflected as equally contingent. On the contrary, the temptation is strong to attribute contingency to certain historical states of society and to call the appropriate observation of these states necessary. Ensuing forms of observation are aptly called "fundamental." Klaus P. Japp, in "The Form of Protest in the New Social Movements," calls this form of observation "protest" and analyzes the paradoxicality, contingency, and continence of this form. New social movements distinguish themselves from society by their way of protesting against it and organize themselves by sorting communication with the help of the distinction "being for" versus "being against." If you are for, say for the protest, you are a member of the social movement; if you are against, you are not. Or if you are for the society, you are against the social movement, and vice versa.[24] The blind spot of the distinctions used by social movements is that the movements themselves, their themes of protest as well as their sorting mechanisms, are part of the very society they protest against; hence they protest against themselves as well. Once again, sociology does not stop at revealing such a paradox of form, but proceeds to ask whether and where the observation of such a paradox actually may take place and how it is handled in furthering or hindering the reproduction of the social phenomenon. In the case of social movements, the mass media were for a long time of great help since they are notoriously unwilling or unable to investigate paradoxes. Instead they prefer conflict, drama, and personality.

There is always a shadow side to all indications. What happens in this shadow makes an important difference. The shadow side is, so to speak, always communicated along with the indication used. It makes a difference, for instance, whether you distinguish humans from gods (and then bemoan their mortality) or humans from animals (and then celebrate their capacity for reason). It makes a further difference if you do not know how to distinguish humans when the shadow becomes populated by computers. Suddenly you do not know any more how to identify humans. Giancarlo Corsi identifies such a shadow side in "The Dark Side of a Career." In modern society the career is one of the most important mechanisms for integrating continual individual biographies with changing societal, above all organizational, demands and opportunities. By searching out and managing discontinuities, a career claims to constitute the continuity of an individual biography. But what are careers distinguished from? They are not distinguished from absent or negative careers, since these are nowadays taken to be careers as well. Take the example of the zero career of the dropout, or the horizontal career of one who frequently changes jobs. Corsi claims that age and aging constitute the shadow side of the distinction of careers. Age means decreasing career opportunities. And it means retranslating the digital (i.e., discontinuous) device of the career into an analog observation of the relationship between individual and society. Consequently, many people try to continue their careers well beyond the point at which they should fade away. In observing their existence within society, they are accustomed to attending to the digital rather than the analog.

Finally, in "The Other Side of Illness," Fritz B. Simon examines how all kinds of observation must not only assure their own capacity to solve problems by offering further operations but also guarantee the existence of something "out there" or "in there" that corresponds to their distinctions. Medicine provides many examples of distinctions that establish themselves in the wake of a distinction between the observable and the unobservable, which doubles as a distinction between the understandable and the nonunderstandable. If you get sick, you are first treated according to these vanguard distinctions, and when you accept them, you are almost bound to accept anything that follows. Simon develops an appropriate notion of the symptom and uses it to describe how symptoms are advanced by a communication system (be it the family of the patient or the physician/patient interaction or even the profession of the physician). The system needs to maintain the distinction between the observable and the unobservable and accept almost everything, depending on the knowledge available, that allows it to maintain that distinction.

Simon's theory of the constitution of symptoms does not preclude therapy; it describes its range and mechanism. The only condition for therapy is that the therapist and physician accept being part of the game. In medicine as in sociology there is a ban on self-exemption. Social-systems theory, by analyzing distinctions as forms, similarly takes part in the game it describes. The essays in this volume are only initial steps in exploring the possibility of such an analysis.[25] Yet a certain research program begins to become visible. It encompasses at least four steps.

First, the analysis must be extended to further phenomena. Distinction theory claims that everything can be analyzed in terms of indication, distinction, and form of distinction. The only condition, but an important one, is that the analysis be situated on the level of second-order observations.

Second, one might usefully inquire into possible distinctions between the notions of "distinction," "difference," "dual" or "duality," *la différance*, Lyotard's *différend* (as he takes on Kant's *Widerstreit*), and so on. There is certainly no point in categorizing these different distinctions, since the question of which term would be used to categorize them would be unanswerable, all of them being perfect candidates. Yet some distinctions would be helpful with respect to (1) whether only one side of the distinction is marked and the other one unmarked; (2) whether both sides of the distinction are marked, thereby pushing the unmarked state into the distinction of the distinction itself; (3) whether the emphasis underlines recursive or iterative changing of the distinction due to repetition, for instance, or instead stresses mathematical identity; (4) whether and how the asymmetry of the distinction is used to build hierarchies;[26] or (5) whether the observer hides inside the distinction or inside the unmarked state.[27] Moreover, many questions relating to previous distinction theories, such as Hegel's dialectics;[28] to the use of negation in observing distinctions;[29] or to the transition from Spencer-Brown's three-valued calculus to many-valued logics[30] remain unsolved. And Spencer-Brown's calculus contains many further ideas that might be fruitful for sociological analysis.[31] Luhmann, for example, recently took up the notions of "oscillator function" and "memory function," which are part of Spencer-Brown's translation of the figure of re-entry into an analysis of time,[32] in order to show how society is proceeding by combining the production of identities with the production of indeterminateness.[33]

Third, it would be interesting to relate the modern use of distinctions in all its different shadings to traditional uses, to compare European distinction devices with non-European ones, or to look into different disciplines' uses of distinctions.[34] Anthropological reliance on distinctions is an interesting case in point.[35] A certain blurring of the distinction between the semantic and the structural will be observed when distinction analyses are extended to different realms of research. The structure of modern society is a structure of semantics, since in it distinctions guide observations that tell us how the world is and how it is not, what we do about it and what we cannot, and what we expect and what we do not.

Fourth, sociological analysis would have to be reconstructed, a prior deconstruction being certainly of help. It would be of great interest to learn not only how general and social systems theories use their distinctions but also how different sociological traditions use theirs. Such a question leads straight into one of the most debated issues of contemporary sociology: the issue of the status of boundaries. Boundaries are distinctions, of course, but very peculiar distinctions, which combine asymmetry with provisions for crossing (or "modern" distinctness with provisions for "postmodern" play). Some sociological traditions other than social-systems theory point to the very precarious status of boundaries of social systems to show the limited range of social-systems theory. Yet the interesting point would be not to show that boundaries never are strict limitations of causal interplay (that is true anyway) but to show how they work as "interfaces" that decouple and re-embed certain domains of an *eigen*-behavior of communication and action.[36]

Further steps could certainly be added to this research program. May it suffice for now to have offered some first ideas.

NIKLAS LUHMANN | **THE PARADOX OF FORM**

I

George Spencer-Brown proposes a concept of form whose implications are hardly exhausted in his *Laws of Form*. Moreover, simply following his form calculus does not readily reveal how sharply this concept of form contradicts traditional thought patterns. The reason for this may be that tradition teaches us to conceptualize form as one side of a distinction whose other side can then be designated in various ways—such as form/matter, form/substance, or form/content. Following this tradition can even lead to the question of what would happen to the concept of form if its conceptual opposite were altered, for instance, if one shifted from form/matter, taken in a cosmological sense, to form/content in its more "artificial" sense relating to works of all kinds. Within this traditional framework, however, how is the unity of the distinction conceptualized, or how can it be conceptualized? This inevitably becomes an issue when form is distinguished from something else. Or to approach it from a different angle: what happens if the other side of the distinction—that is, matter, substance, or content—is simply omitted from consideration and form as such becomes the main focus of reflection and manipulation? What happens if "formal logic" is pursued as an attempt to ground all distinctions in themselves—an endeavor Gödel eventually halted? What happens when we raise the Derridian question of what the form, as a precondition for being form, makes appear as "present,"[1] and furthermore what is thereby excluded from consideration? And what happens, if one contests that signs have a reference to objects, and that signs are only allowed to partake in the play of differences among themselves? Does difference then become the ultimate form, whose unity can only ever be designated by constantly new differences?[2]

Drawing a distinction is an operation that remains possible even when the distinction is observed in its form as a paradox. It can be done when and as long as the "autopoiesis" of observation—and likewise the autopoiesis of life—works. A living organism consists of nutrients previously taken in; this insight, however, need not prevent the organism from caring about its future subsistence. Consciousness operates on the basis of former thoughts; however, this insight does not lead to the cessation of thought. Operating and observing therefore have to be distinguished even though making this distinction in turn is an operation of observation. At this point, theory is forced to perform an autological inference, that is, an inference back to itself. This means that everything we do or do not observe at all is founded on a paradox and, at the same time, that the distinction leading out of this paradox is the very distinction between operation and observation.

II

The form of the form is for Spencer-Brown a distinction and consequently a condition of possibility for observation, which naturally presupposes further conditions if observation is to be possible as an *operation*. A form is thus something with two sides that are distinguished. The clarifying passage preceding all definitions and positing of axioms reads: "We take as given the idea of distinction and the idea of indication, and that we cannot make an indication without drawing a distinction. We take, therefore, the form of the distinction for the form."[3] Why "therefore"? This question will be the focus of the following considerations.

A form has two sides; so much seems to be certain. A form is established by fixing a boundary. This leads to the separation of two sides such that getting from one side to the other requires another operation to cross the boundary.[4] Establishing a form is thus distinguishing, and distinguishing is an operation that, like any operation, requires time.

What *is* a form? Note that this question is carefully avoided. In any event, form cannot be something that makes something else appear as "present" (to allude to Heidegger and Derrida once again). The operation can either take place or not. The present is thus not a stretch of time in which, by means of a division of time into past, present, and future, a dynamic entity exists. Instead the present is the boundary that is drawn when it is important to delineate past and future as different. Hence the present is not an occasion for "presence" to occur but only an occasion to observe the difference between things past and future.

Form consequently does not possess any ontological status. We only state here what is meant when form is employed. Form not only *is* the boundary, but also contains the two sides it separates. Form has, as it were, an open reference to the world; this might underlie Spencer-Brown's enigmatic statement "Distinction is perfect continence" (p. 1). Are we to understand "perfect continence" to say that the distinction contains itself as well? How can the distinction be perfect otherwise? How else could the distinction divide the world into two sides, a process that can only happen within the world?

With these questions in mind we turn anew to Spencer-Brown's opening passage. Each form has two sides that it distinguishes. Those two sides, however, do not participate in the shaping of the form in an equal manner. The operational use of the form can only proceed from one of its two sides. The operational use has to begin somewhere, for it would otherwise be unnecessary to distinguish the two sides at all. The distinction is made with the pragmatic intent[5] to designate one side but not the other. What is distinguished, therefore, has to be distinguished from the distinction itself. Such a formulation might be dismissed as a mere rhetorical trifle; Spencer-Brown avoids this by differentiating the terms "indication" (i.e., signification, designation) and "distinction." Yet, with or without this terminology, the problem remains. We cannot begin with the operation as long as the distinction between distinction and indication (signification) is not copied into the distinction.

Is the distinction that recurs in itself, and that cannot occur in any other way, then, the same or not the same distinction? Toward the end of his elaborated calculus, Spencer-Brown addresses this problem as that of the "re-entry" of the form into its own space, thus the re-entry of the form into the form, of the distinction into what is distinguished. The problem is present from the beginning,[6] and in the end it receives its own form. As is the beginning, so is the end situated outside the calculus. The problem now has a name, a designation. But this is just a form of self-reference, thus a kind of return to the problem. Says Louis H. Kauffman, "A mark or sign intended as an indicator is self-referential."[7] In any event, a mark is nothing but a form distinguished from the distinction.[8] Thus the problem lies in the unity of the form—a form that only occurs if it is used as the condition of possibility for operations.

III

As stated previously, only the one side of the form, that is, the designated side, can be used operationally. Using both sides at the same time would infringe upon the purpose of the distinction. It is not possible; it would result in a paradox since one would have to call at once what is different the same. We cannot help but accept that the form of the form is a paradox. Thus the identity of a difference, that is, a distinction that distinguishes itself in itself, is the issue. And saying "a distinction that distinguishes itself in itself" could lead us to surmise that we are dealing here with a symbol for the world. Observed as a paradox, *any* form symbolizes the world. As is the world, so is the paradox a case of pure self-reference—not merely a vacillation of opinion but moreover the fascination effected by this vacillation. A paradox is a form that contains itself without any reference to an external standpoint from which the paradox could be observed. The paradox is hence both beginning and end at once. However, the observer is a system who continues his operations. By making the transition to another distinction, he disassociates himself from the paradox. Still, if it is a paradox, how can it be dissolved at all? How can it be unfolded; that is to say, how can it be transformed back into stable, discernible identities?

Spencer-Brown's initial command, "draw a distinction" (p. 3), provides a first clue. A command can be either carried out or not; that is its form. If the command is carried out, then the calculus of form can be operationally performed; if not, then not. And it would not be an exaggeration to add that only if the command is carried out can one observe[9]—otherwise not.

Dissolving the paradox into the form of a command brings time into play. The operations of the calculus (as well as observation *per se*) have to be performed sequentially. Spencer-Brown lets the mark march, step by step. The mark is repulsed and attracted by the paradox of the re-entry, as it were, and the world becomes ordered in this interplay of repulsion and attraction. Beginning and end are the same, and not the same; and in between (or: in the meantime) the world achieves its organized complexity. In retrospect it becomes clear that the initial paradox of form already contained a time paradox. The distinction is only a distinction if it provides both sides simultaneously, but the operations and especially the crossing of the boundary back and forth can only be performed sequentially. From a structural point of view, the two-sided form only exists in the temporal mode of simultaneity; operationally considered, however, the two-sided form can only be actualized in consecutive operations since operations that proceed from one side exclude operations that proceed from the other side. The form is the simultaneity of sequentiality.[10]

But there is still a third way to observe the dissolution of the paradox of the form, namely by asking: Who is Spencer-Brown? Who is the one who tailors all this in a Boolean fashion? Who tells the story, and does the narrator appear in the narration?[11] The observer is Spencer-Brown himself, who wants to force us, through the strict form of calculus, to perform the calculus the same way he does, and thus not to distinguish between different observers.[12]

In the traditional, ontological worldview, observers were indeed forced to see the same and thus to be identical observers. The other side of the form of observation was regarded as error. By crossing and recrossing this boundary, one could assure oneself of this view and eliminate the erroneous opinion. It was assumed that repeating the procedure brings about the same result. The transcendental-theoretical critique of metaphysics still makes the same presupposition, which, however, is now located in the systems of consciousness and framed as *a priori* conditions of their (identical) ability to function. This point of convergence of the systems of consciousness is then named *subject*. Despite all doubts about the ontology of the world (the term "ontology" dates back to the seventeenth century and was even then a symptom of incipient doubt), the calculus of form finally seems to be the ultimate form in which observers are still able to ensure their agreement. But how is that possible, given that the calculus only leads from paradox to paradox?

It may help to define observation as the use of a distinction for the purpose of designating the one side (and not the other), however this is achieved—whether by making use of consciousness, or through communication, or through a programmed computer. Accordingly, Spencer-Brown states: "An observer, since he distinguishes the space he occupies, is also a mark" (p. 76). Operationally speaking, an observer emerges as a system through a consecutive sequence of his observational operations. Since the use of distinctions for designating the one (and not the other) side of the form can be observed in turn if attention is focused on this and not on something else, it is possible to observe observers.[13] For this to happen, one must be able to distinguish observers (distinctions), including oneself as an observer at different points in time. The paradox of form can then be unfolded by identifying different observers without necessarily, in fact without possibly, leading to the assumption of "intersubjective" agreement. Different observers cut through the world in different ways, distinguish differently, use different forms, and thus construct the world not as a universe but as a "multiverse."[14] How communication is then possible remains, it is true, an open question for traditional subjectivism. But our answer can now read: through communication, that is to say, through the formation of a system of observation *sui generis*, that is, through the formation of social systems.

Hence there are different possibilities for dissolving the paradox of form: factually by means of a directive that may or may not be obeyed; temporally by the sequencing of operations that are bound by the fact that they always carry along another side, thus a simultaneously effective distinction; and socially by distinguishing different observers who all base their observations on different distinctions. On the level of second-order observation it is then possible to see and state that there are different stopgaps, and furthermore that the paradox of form only comes about because an observer attempts to observe at once both the unity and the distinguished sides of a distinction. A second-order observer can thus see that another observer runs into this problem when attempting to observe how (i.e., with which distinction) he is observing. The second-order observer, mind you, is a first-order observer as well, for he must distinguish and designate the observer he intends to observe. Second-order observation thus does not rid us of the problem. It only confirms that we are dealing here with a universal problem, one with which every observation is confronted.

IV

One of Spencer-Brown's many self-commentaries also hints at this conclusion. According to Spencer-Brown and many others, the world seems to be suited for self-observation. To this end the world created physicists who have to distinguish themselves as observers from what they observe. Physicists can only do this if they operate self-referentially.[15] That physicists (the notion includes both people and instruments) especially are competent to do this has become evident only in our century. It is true that August Wilhelm Schlegel had already touched on this thought; however, he rejected it as impossible and nonsensical.[16] In any event, the observer (in a world that uses him for observing itself) positions himself on one side of the form called observing. But then what about the unity of this distinction, that is, the self-observing world?

The tradition always started out from an existing, present world that, to be sure, confronted humans with problems of understanding, knowledge, and ability. In Milton's *Paradise Lost*, the archangel Raphael explains world history to Adam (i.e., to the reader) amidst an already ongoing world history. The explanation concerns the paradox of eternity and evanescence and, against this backdrop, the meaning of God's game with the devil. That world history can only be explained within world history corresponds to the observer's paradox of form and thus to the necessity of distinguishing and, from a temporal point of view, of limiting operations to a present that is distinct from its past and its future. Spencer-Brown dissolves this paradox with the help of the observable/unobservable distinction, which of course in turn presupposes an observer who handles the distinction and who is forced by the distinction to refer back to himself. "We may take it that the world undoubtedly is itself (i.e., is indistinct from itself), but, in any attempt to see itself as an object, it must, equally undoubtedly, act so as to make itself distinct from, and therefore false to, itself. In this condition it will always partially elude itself" (p. 105).

In the Christian tradition, equating striving for knowledge with the Fall of Man may count as a sufficient explication of the problem. Correspondingly, in Sufic mysticism it is the doctrine of the fall of the angel Iblis, who sees through the paradoxical nature of God's command—that he bow to Adam—and dissolves the paradox in his own way, namely by disobedience.[17] But since God always knew and intended all this, Spencer-Brown can only take it as a practical joke[18] that may have served to familiarize humans with the creation enterprise's paradox of form.

In earlier societies one would always distinguish between a familiar and an unfamiliar domain of the world (as the one and the other side of the world's form), and then would at most attempt a re-entry, that is, a symbolization of what is unfamiliar within what is familiar. Highly developed religions obviously point to more than just this. Hence it may not be completely off track to surmise that the paradox of form is today a convincing possibility for symbolizing the world within the world (and not only for symbolizing what is unfamiliar within what is familiar).

Within this framing of the "meta-physics" problem, an observer, trying to unblock his observing, cannot count any longer on ontological tools. He also no longer considers himself a subject who can still hope to find within himself, or in reflecting on language usage, foundations for agreement with others. The self-paradoxification of all observing without exception (and God is not the exception but the principle) leads to the fact that the observer can no longer know where he stands, but can very well know how he moves. The observer is left with only the facticity of his individuality, the way he obeys the instruction: draw a distinction.

V

History makes this possible: a more recent theory can compare itself to an older one, while the reverse is not possible. We use this advantage of contemporary theory. We already mentioned the theological tradition and "subjectology" with regard to the genealogy of the observer. Along with this genealogy, however, what is observed has changed as well. This can be demonstrated by an important, maybe the most important, observational tool of the old-European tradition. We shall call it the *genos* technique, which enables one to comprehend the world in the form of classifications.

As far as records go, the tradition begins with Plato's *Sophist*, and here, from the outset, the defense against paradoxes is still explicitly stated. The art of distinguishing[19] (or should one say, the art of grasping) called *diairesis* is brought into a particular form of "dialectics." This art is possible because genera and species are distinguished. Thus, the possibility of thinking that the same could be something different or something different could be the same is excluded,[20] *even though "genos" designates a general idea permitting obviously different manifestations at the level of individuals.* Thus *diairesis* is obviously viewed as a strictly speculative, visionary synthesis of different things into one group that can be distinguished. And this is even offered as a science (*epistēmē*) for free people; one could almost say: as a civic science.

This *genos* technique determined the tradition through and through. It was probably applied wherever rationalization was concerned, for example in the categorization of evidence for decision in Roman civil jurisprudence.[21] Realism as well as nominalism made use of it and staged their controversy on this basis. The paradox then migrated to rhetoric[22] and poetry.[23] But even here the *genos* abstraction remained the material for play with which the paradox was lured out of hiding. Kant perceived it for the first time as nothing but a historical device that no longer gives "noticeable pleasure" but can only be handled as a cognitive technique.[24] As a result, Kant could use the now vacant concept of dialectics differently by employing time as a schema for asymmetrization.

Consequently, the art of distinguishing, that is, the *diairesis* of free people, is set free as well. No longer is it tied to worldly classifications that can be recollected as ideas. Hence it also is no longer bound to prerequisite "categories" in the Aristotelian sense. And the *diairesis* cannot at all be handled in terms of the old concept of *epistēmē*. But what can one say *instead*? With what could one start *instead*, once decomposing Being into categories no longer yields "noticeable pleasure"? Perhaps with a self-referential dissolution of the paradox that the same is something different, and something different is the same?

There is no standardized prescriptions any more for how this has to happen, not even in the sciences. Everyone has to cope with his own living situation on his own. One does not even have to convince oneself or others. It may or may not succeed. There are at least interconnections. Nowadays, for example, the stock exchange, where options on unfolding the paradox are traded, is called "culture."

VI

The late Renaissance rhetorical-poetic technique of paradoxification saw its purpose in stimulating thought by making uncommon proposals, which did not necessarily mean getting stuck in the impasse of the paradox. At first, there is nothing strange in the suggestion that one search for alternatives to the "comun parere," to commonly accepted understandings. Proposing uncommon explanations can be regarded outright as the program of scientific innovation. As the climax of his detailed critique of logical empiricism in social psychology, Kenneth Gergen demands, for example, "generative theories," and for this he recommends a procedure that four hundred years ago would have been subsumed under the concept of paradox: "One may also foster generative theory by searching for an intelligent antithesis to commonly accepted understandings."[25] Such formulas, however, are offered today in science, which has forgotten Europe's intellectual history and which is therefore compelled to repeat them in the form of new discoveries. This, one could achieve more easily.

But: "the same" is different nonetheless, and historical orientation can only mean *not* to do something as it used to be done. An element of negation must be built into the historical self-reference of a semantic discourse. Only in that way, that is, by means of self-paradoxification, can this also be related back to the principle of surprise in scientific discoveries and generative theories.

At least by the eighteenth century, the sense for communicating paradoxes either had gotten lost or had been replaced by a taste for parodying the generating of paradoxes. Covered by the new state political order, the new self-certainty of knowing gained prevalence and, as a sort of background certainty, enabled a tendency, bordering on naïveté, toward enlightening and moralizing. Even today paradoxes are regarded as nothing more than a logical and, by means of logic, avoidable (or excludable) problem. The rhetorical tradition is not remembered anymore.[26] This should not be taken as saying that we could turn back the clock. We *differ* from the skeptical orientations of the sixteenth-century late Renaissance—though they are commended as examples today[27]—at least in two respects, which can be determined right down to the last detail: (1) For us, the objective of a science free of paradoxes, admitting only clear and distinct elements and concepts, has to be abandoned *since extracting those elements and concepts to be admitted requires a nonreflective use of distinctions.* (2) The way the late Renaissance generated paradoxes via puns must be abandoned likewise *since we lack the trust in words, even in the rhetoric-semantic distinction between "verba" and "res."*

In today's classic understanding of science the distinction between level and meta-level serves as a form of unfolding paradoxes; provided that one does not address the unity of this distinction, science's freedom from paradoxes seems to be guaranteed. In functionalism,[28] colliding with the aforementioned classic understanding, it was the distinction between problem (what the problem concerns) and problem solving (functional equivalents) that was postulated in the ratio of one to more-than-one, although problem and problem solving cannot be identified independently of each other. When one asks subsequently for the unity (or the shared premise) of this controversy, or when one poses the questions in the typical sociological fashion—what is behind all this?—one then comes across the observer, or more precisely, the necessarily paradox-laden form of observing by means of distinctions and indications. One can then see that the controversy is about different ways of unfolding paradoxes, or about distinguishing distinctions. Posing the question of unity would lead to the paradox of the identity of what is different *and must therefore be avoided*.

It is self-explanatory that the distinction itself between paradox and unfolding is paradoxical. But this just substantiates once again the insight that we are dealing with a self-referential concept replacing that of ultimate reasons—perhaps even with the concept that will supersede the notion of Reason as being inherently reasonable.

Spencer-Brown has good reasons to reject the subsumption of his calculus of form under the domain of logic, since logic is governed by truth. Paradoxes, however, are not a form of truth, hence also not a form of untruth. Paradoxes were once called monsters of truth;[29] better yet, they have been said to tap into the font of true knowledge.[30] The concept of the paradox of form now provides a better reason why paradoxes are not a form of truth. We repeat: Any observation presupposes a distinction whose unity can only be designated paradoxically. Observing, therefore, cannot (or can only paradoxically) designate itself. Or it has to use *another* observation to which the same then applies. The truth/untruth code is just one of countless possible distinctions, just one of many forms that owe their unity to their two-sidedness.

Whether an observer places importance on truth is thus an open question answerable only by observing the observer; since the sixteenth century, the theory of art at least has rebelled against this truth precept. Above all, the concept of paradox cannot therefore be defined by basing the definition on the concept of logical contradiction. This need not imply that one has to go back to the weak rhetorical concept of paradox. This might have been perceived as a provocation (and, thus, as meaningful) in the world of the sixteenth century. Today something as easily obtainable as a deviation from what is usual does not suffice as a foundational concept. To bind the concept of the paradox to the necessary use of form in any observation is therefore recommended, a course leading to the thesis that there cannot be an observation in which self-designation would not be blocked by a paradox. Observing is a paradoxical operation. And only because of that can this operation potentially designate truths, but precisely only in contrast to untruths and without considering the unity of the very difference between true and untrue.

With all this said, an essential question remains unanswered: If logic cannot claim ultimate competence for the paradox of form, then perhaps religion could? Does the interest in paradoxes, so conspicuously increasing in intellectual and, above all, in philosophical movements of this century, run parallel to a surprising viability of religion in a society that has understood itself to be "secularized"? That may be so, but then this raises questions for the traditional corpus of religious doctrine. Or, to conclude with a voice from the seventeenth century: "Methinks there be not impossibilities enough for an active faith."[31]

DAVID ROBERTS | **SELF-REFERENCE IN LITERATURE**

Introduction

In the short history of the reception of Niklas Luhmann's systems theory by literary scholars, two responses can be discerned. The one—represented by Siegfried J. Schmidt's book *Die Selbstorganisation des Sozialsystems Literatur im achtzehnten Jahrhundert* (1989)—analyzes, as the title indicates, the emergence in Germany in the second half of the eighteenth century of an autonomous or self-organizing literary system in the course of and as part of the process of the functional differentiation of society. The interest of this study, which "systematizes" in a double sense Schmidt's earlier project of an empirical theory of literature, lies in the constitution of literature as a social system and accordingly in a switch from a hermeneutics of the text to a sociology of the literary system. The other line of reception—represented by Dietrich Schwanitz (*Systemtheorie und Literatur*, 1990)—takes as its organizing theme the self-reference of literature, which is seen as operating by means of a difference in information between the observer and the observed. Here the self-organization of literature is approached not through the sociohistorical process of functional differentiation but in terms of observation and self-observation. The two perspectives are complementary since functional differentiation is dependent on the feedback given by self-observation. The underlying difference is nevertheless significant. Schmidt's formula for paradigm change—from text to literary system—is directed to a new conceptual framework for the sociology of the literary system, whereas Schwanitz's "new paradigm" is oriented to the evolution and differentiation of literary forms (genres).

The focus of the present paper—literature and self-reference—is close to that of Schwanitz. That is to say, my interest is the internal mechanisms of literary self-differentiation, which may in general be taken to correlate with the whole social process of functional differentiation, but which can be analyzed in their own right. The immediate stimulus of my paper can be found in two recent essays by Niklas Luhmann in which he develops in a highly original way the implications of Spencer-Brown's *Laws of Form* in relation to European rationality and to modern art.[1] I follow here Luhmann's reading of Spencer-Brown in my attempt to explore the implications of the "laws of form" for literature.

The Laws of Form

The Laws of Form is an inquiry into the paradox involved in all observation. For the world to see itself

> it must first cut itself into at least one state which sees, and at least one other state which is seen. . . . whatever it sees is *only partially* itself. We may take it that the world undoubtedly is itself (i.e. indistinct from itself), but, in any attempt to see itself as an object, it must, equally undoubtedly, act so as to make itself distinct from, and therefore false to, itself. (p. 105)

The world makes itself distinct from itself by means of a distinction. "We take as given the idea of distinction and the idea of indication and that we cannot make an indication without drawing a distinction" (p. 76). Drawing a distinction severs the world into a form with two sides. On the one side, the inside, is the indicated or marked world, on the other side is everything else, that is, the unmarked world. To observe is to draw a distinction; that is to say, observation comprises the operative unity of distinction and indication. Observation presupposes, however, an observer, but where, after having drawn a distinction, are we to place the observer? The observer is not distinct from the distinction. He cannot as yet be placed on either side of the distinction. To decide where he belongs requires a second observer. In other words: observation uses a distinction but cannot distinguish the distinction it uses. The operative unity of distinction and indication conceals a second distinction, the distinction between distinction and indication; that is, the distinction between the observer and the observed. The introduction of this second distinction (second observer) is what Spencer-Brown calls "re-entry into the form." It allows him to conclude: "The first distinction, the mark, and the observer are not only interchangeable, but, in the form, identical" (p. 76).

That is, the world marked or indicated by the first distinction is identical with the observer. Only a second distinction can distinguish the first distinction and thereby the observer who uses it, distinguish that is, between the observer and the operation of observation.

The first distinction (observation) divides the world into a form with two sides, the seen and the unseen. The second distinction (observation)—the re-entry of the form into the form, of the distinction into the distinction—divides the seen in turn into the seen and the unseen, that is, the observed and the observer. The first observer is visible only to a second observer. Observation is blind to itself and cannot distinguish between itself and what is observed. Nevertheless, the observer is of course invisibly present. As Luhmann puts it, every distinction presupposes itself and thus excludes itself from what it can distinguish.[2] Thus what the second observer makes visible is the invisible observer and the invisible re-entry into the form which is the presupposition of all observation. This presupposition creates the asymmetrical difference between the two sides of the form. There is no way back to the lost unity, the unmarked or formless state of the world. The world can be observed only by means of distinctions that make the unmarked world invisible. What the observer observes is the marked world, the indication made by drawing a distinction. The unmarked world, the world as it really is, however it is, is neither observable nor describable. Form is two-sided, but the observer cannot get to the other side of the form. The observed world is the product of the observer. This asymmetry is the condition of an increase in the complexity of observation of the observing system and of the operative closure of observation that observes by means of distinctions internal to the system.

Observation makes the world both visible (marked state) and invisible (unmarked state). It also makes the observer invisible. The observer cannot see himself. There is always a blind spot to observation. To repeat: the re-entry of form is invisible. It cannot be observed since it has already occurred through drawing a distinction. It is the invisible presupposition of observation. Thus the re-entry of form can only be observed by a second observer. The first observer cannot see himself. He can only be seen by a second observer, who observes the first observer by means of the distinction between distinction and indication. In turn the second observer cannot see himself.

We can thus distinguish between first-, second-, and third-order observation, that is, between observation, observation of observation, observation of observation of observation, and so on. For first-order observation the re-entry of form is invisible, for second-order observation the second-order re-entry of form is invisible, and so on. Thus if the invisible re-entry of form is the presupposition of first-order observation, the invisible second-order re-entry of form is the presupposition of making first-order re-entry visible. Only with second-order observation does the difference in first-order observation between the inside and outside of form become visible as the difference between inside and outside *inside* the form. The second-order observer can thus distinguish between distinction and indication, that is, between reference to the observer and reference to the world, and observe by means of this distinction. For the first-order observer by contrast this distinction (the re-entry of form) is invisible. He cannot distinguish between self-reference and external reference. First-order observation cannot see itself and thus lacks self-reference. Or more exactly, since (the invisible) re-entry of form is the presupposition of observing the world, self-reference remains invisible (latent), and external reference alone is visible (manifest). The first-order observer sees the world and not himself observing the world. The observer disappears into the observed world and does not comprehend himself as observer in the world.

First-order observation sees the world; second-order observation sees the observer in the world observing the world. Each observer has a blind spot. The infinite regress thereby opened up can only be halted or, perhaps better, contained, by *self*-observation. Self-observation is second-order observation that operates by means of the distinction between self-reference and external reference. From this it follows that the self-reference of self-observation is defined by the difference between self-reference and external reference (observe here the re-entry of form). The self-reference of self-observation (as opposed to the self-reference observed by a second observer) I shall call self-reflection, since it involves the difference between observation of observation from outside and self-observation from inside.

Spencer-Brown defines form as two-sided, as the unity of a distinction. His starting point is thus difference and not unity. It enables him to show and us to see that first-order observation uses difference but cannot see this difference. The first-order observer does not see the difference between the unmarked and the marked world—he thinks he sees the world, just as he uses but cannot see the difference between self and world.

The fact that all observation of the world originates in difference makes all the difference. Thus for Niklas Luhmann it signals the break with all conceptions of the world that are based on unity. It is the key to his critique of European rationality and the means whereby he distinguishes systems theory as a new universal theory from the tradition of European rationality, proposing instead a concept of system rationality based on a theory of difference. Luhmann's argument is this: the problem of the re-entry of form remains latent in the ontology of premodern rationality. The assumption of a rationality continuum between thought and being joins observer and world; for example, being is represented in thought, nature is imitated in art. This corresponds to first-order observation and might be called the one-world vision of the world. The transition from traditional to modern society, from stratified to functionally differentiated society, brings about the dissolution of this rationality continuum. Whether reason is now divided into different types—instrumental or value rationality for instance—or whether the alternative response of holism is taken—as for instance in German idealism—Luhmann sees the re-entry of form as posing the core problem of European rationality. Thus distinguishing various types of rationality does not ask what the unity of the distinctions employed is; holism in turn uses as its differential schema the difference between the whole and the part. Re-entry thus operates on the assumption that the part represents the whole. Or to put it more abstractly: holism is based on the identity of difference and identity. Systems theory, by contrast, is based on the difference between identity and difference. In other words, systems theory understands itself as a system observing the world in the world and not as the part that is at the center of the whole.

The question systems theory asks of European rationality is, where is the observer? As we have seen, this is a question of form: not what is distinguished (observed), but who distinguishes and how. This blind spot of rationality points to the need for an epistemology based on the observation of observation (second-order cybernetics). It offers a way, Luhmann suggests, of coming to terms with the modern plurality of discourses, and the attendant issues of "relativism," "historicism," "deconstruction," and "anything goes." Systems theory has no problem with the multiplicity of discourses; they are the necessary consequence of the ongoing functional differentiation of society and the corresponding increase of complexity. What Luhmann objects to is what he calls the laziest of compromises—"pluralism." The example he takes comes from reader-reception theory: every reader is granted his own point of view and interpretation but only within the space allowed by the text, which remains the "objective" world. Similarly, recent philosophy of science grants theory the epistemological status of "constructivism" but qualifies it by reference to reality. By half-heartedly clinging to a subject/object distinction these "solutions" half acknowledge and half deny the implications of the re-entry of form.

The re-entry of form into the form is a self-implicating form. Systems theory accordingly is a self-implicating theory, which starts from the distinction system/environment and uses *this* distinction to observe the world. A self-implicating system requires autological concepts by which the observer recognizes himself as a system-in-an-environment. (Here, to avoid misunderstanding, it must be stressed that "observer" is a convenient abbreviation for an observing system, whether personal or legal, economic, educational, medical, etc.) That is to say, re-entry involves the re-entry of the distinction system/environment into the system, which takes the form of the distinction between self- and external reference. Autological concepts suppose second-order self-observation by means of which the observer observes how he divides the world through the difference between self- and external reference. The outcome is a "constructivist" view of the world in which the unity of the world and its determinability through observation are no longer congruent. This leaves us with the certainty, as Luhmann expresses it, that observation makes the world both visible and invisible.

For Luhmann the observation of first- or second-order re-entry amounts to the observation of the production and unfolding of a paradox. To some extent the paradox can be dissolved by means of a distinction, which Luhmann tends to neglect for the sake of paradox. If we distinguish between the unmarked and the marked world (World and world), then the paradox of observation making the world visible and invisible disappears. Equally, we need to distinguish, as Luhmann does not, between two-sided form (Form) and the forms that are produced on the one side (form). Two-sided form is the unity of a distinction, but all the forms that are produced by drawing a distinction appear on the one side of the form, and this one-sidedness or asymmetry produces and is produced by the re-entry of form.

These distinctions are necessary in order to bring out more clearly the paradox of form, to which forms, since they exist, are always the "solution." The paradox is this: what is outside the form is accessible only inside the form. The world observed is always the world in form. Luhmann thus arrives at his concept of system rationality. A system separates itself from its environment and achieves operative closure by means of the difference between self- and external reference. This difference makes the system indifferent to its environment and serves as the basis for the elaboration of its own complexity. Rationality is thus to be understood as the ongoing self-production or autopoiesis of a system by means of its own operations, which elaborate its ever-more-improbable complexity. Functional systems differentiate themselves through the self-referentiality of second-order observation. Such systems are the operative executors of the rationality of modern, that is, functionally differentiated, society. Their second-order observation does not, however, overcome the blindness of rationalism. All second-order observation can do is make this blindness apparent.

How does this apply to art? Luhmann proposes in his essay "World Art" that the difference between first- and second-order observation distinguishes modern from premodern art. Traditional or premodern art is characterized by the mimetic relation to the world of first-order observation. Traditional art accordingly is described as object art. Modern art is characterized as world art—a title intended to bring out the paradox of the situation in which art has lost the certainties afforded by mimesis, the "one world" that unites observer and world.

The loss of the unity of the world is thus the precondition of "world art." The unity of the world, that is, the world in its unmarked state, has become the transcendental pregiven of all artistic observation. It can be thought of not as the object but as the medium of art, a medium that is given form by the self-posited forms (constructions) of art. Only on this level does art become universal, that is to say, liberated from dependence on objects. In other words, aesthetic experience, that is, aesthetic observation, because it no longer depends on objects, is universally applicable and realizable. The price of this independence from objects, however, is a dependence on constructions that cannot exclude other possibilities. As with all functionally differentiated systems, universality takes the form of modality, the form, that is, of an ever-present choice of possibilities. Thus the constructions of world art can only prove themselves by their capacity—or by their failure—to carry contingency. The World itself is not contingent. Contingency is the product of the differentiation of systems.

The transformations of the world from the object into the medium of art is brought about by the difference between first- and second-order observation. I stress the difference between the two because second-order observation—whether external or self-observation—takes first-order observation of the world as its object, thereby transforming the observed world into its material or medium. This means that second-order observation uses but cannot replace first-order observation. Second-order observation not only introduces a distancing difference into observation, it lives from this difference. World art therefore cannot be purely self-referential. Second-order observation involves partial self-reference that does not determine the system as a whole. Aesthetic observation, as we saw, requires distance or liberation from dependence on objects; aesthetic experience in the modern sense can thus be formulated as the consciousness in the observer of the difference between first- and second-order observation in the work of art. The traditional formula for the unity of this difference is form and content.

Since the first-order observation of object art makes the observer invisible, only the re-entry of form in second-order observation makes the self-reference of art visible, that is, its liberation from dependence on objects, or again, its formal autonomy. The term "world art" thus refers to the autonomy of art in relation to the world. The difference between first- and second-order observation is the key to the autonomy of art. It compensates for the inaccessibility of the world, the loss of the "one world" of premodern art. Because the world can no longer be observed as unity, the observation of observers takes its place, and this is the recursive operation that brings about the differentiation of a functional system of art, that is, a system of communication by and about art that organizes its self-observation by means of a temporal schema. Innovation—the difference between the old and the new—becomes the program of the self-production (autopoiesis) of the autonomous system. The autonomy of art has two aspects—social and world autonomy. Art differentiates itself from the rest of society as a separate self-organizing social system (involving production, reproduction, reception, criticism, etc.), and this is the social corollary of the world or aesthetic autonomy that recognizes that works of art observe the world according to their own internal self-referential criteria.[3] The attainment of social and world autonomy signifies the uncoupling of art from the premodern understanding of art as social representation or as imitation of nature and from all extra-aesthetic demands (whether religious, political, moral, didactic, etc.).

This process of the differentiation of the system of art stretches from the sixteenth to the end of the eighteenth century and is part of the whole process of the transformation of stratified society into a functionally differentiated social system. A third level of observation now appears—aesthetic theory, which seeks to determine both the specificity of art, that is, the idea or concept of art, and the historicity of art. The idea of art, which emerges as the theoretical reflection of the social autonomy of art, serves as the unifying focus for the observation of the art of the past (which becomes thereby "art" in the modern sense) and for the determination of the identity and difference between premodern and modern art.

For Luhmann, however, third-order observation of a functional system has a further task: to determine the unity of the system in terms of its coding, which is scarcely conscious on the level of first- or second-order observation. By coding, Luhmann understands a binary code that defines the constitutive difference organizing the unity of the system. Strictly speaking, this constitutive difference is the (unity of the) distinction by which theory observes the operations of the system. What is this code in the case of art? The traditional answer was the difference beautiful/ugly. Eighteenth-century aesthetic theory proposed the difference between the general and the particular. Luhmann's answer takes the form of a question: "what could the code of art be, if it does not already lie in the form that allows it to distinguish self-reference and external reference."[4]

Is this question/answer tautologous, or is it the empty form that is given life by the re-entry of form? In the second half of this paper I want to apply the laws of form to literature—"literature" understood as a concept that is a product of the functional differentiation of art and that is now used to cover both premodern and modern literature.

The Self-Reference of Literature—Romantic Irony

The intense burst of aesthetic theory around 1800 which we find in German romanticism indicates the emergence of a new level of observation of the literature of the past and present, whose crucial element is the consciousness of the difference between first- and second-order observation. In the theory of romantic irony the laws of form become the key to the concept or idea of art (literature). And, as Walter Benjamin has shown in his doctoral dissertation *Der Begriff der Kunstkritik in der deutschen Romantik*, published in 1920, the key that unlocks the theory of romantic irony is Fichte's definition of reflection. We are already familiar with this definition since what we find formulated here is the operation of the re-entry of form into the form: "The act of freedom, by which form becomes the form of the form as its content and returns to itself, is called reflection."[5]

Romantic irony we can thus think of as the point at which the paradox of (literary) form is raised to theoretical consciousness, or alternatively as the point at which literature comes to self-consciousness. In romantic irony critique and creation, theory and practice fuse to give, as the French philosophers Philippe Lacoue-Labarthe and Jean-Luc Nancy put it, the moment of the invention of modern literature.[6] (It is precisely this moment that signified for Hegel the end of art.)

Romantic irony is thus the theory of (modern) literature in the form of a theory of the self-reference of literature, the key to which can now be comprehended thanks to Spencer-Brown as lying in the re-entry of form. Before turning, however, to the practice—or, if you prefer, the practical theory of—literature's self-reference, I want to remain with Benjamin because he brings out particularly clearly the paradox of form as it is presented in romantic irony, but also because Luhmann explicitly acknowledges and distinguishes his own concept of world art from Benjamin's reconstruction of the German Romantics' theory of the work of art and the idea of art.[7] Nevertheless, for all the evident difference between systems theory and German idealism, there are some surprising parallels in their treatment of the paradox of form, which are perhaps not so surprising now that we have discovered their common ground in Fichte's concept of reflection.

The vantage point of observation in the romantic theory of literature is given by the concept or idea of art. The idea of art is an absolute concept (what Lacoue-Labarthe and Nancy call the "literary absolute"), and it is the function of form to point beyond its own finitude to the unattainable idea of art to which the individual work strives. Form is thus the key to the romantic theory of the work of art. As the self-limitation of the infinite energy of reflection, form has two sides: the individual work points beyond itself to the absolute work of art. We thus have a mystical concept of two-sided form: beyond the limits of the visible work lies the invisible work. Visible form is therefore necessarily contingent, and it can only point beyond itself by making its contingency apparent. This is the task of romantic irony, which we can reformulate in terms of the paradoxical achievement of the re-entry of form into the form. Romantic irony is the objective self-reflection of the work because it does two opposed things at the same time: it manifests the self-limitation of the form within the form at the same time as it raises the form to a higher power, the power of potentially infinite self-reflection. This paradox of form that is both less and more than itself we might call the mystery of two-sided form. It is, nevertheless, the logical consequence of the laws of form, that is, of the initial distinction that divides the (unmarked) world into observer and observed. I quote Gotthard Günther:

> It stands to reason that these systems of self-reflection with centers of their own could not behave as they do unless they are capable of "drawing a line" between themselves and their environment. We repeat that this is something the Universe as a totality cannot do. It leads to the surprising conclusion that *parts of the Universe have a higher reflective power than the whole of it.*[8]

The re-entry of form thus demonstrates simultaneously the self-limitation and self-potentiation of the part in relation to the whole.

We are now in a position to understand the paradoxical dialectic of form in romantic irony. Benjamin distinguishes between subjective and objective irony in terms of (the unity of) the difference between self- and external reference. Subjective irony ironizes content; objective irony ironizes form. Objective irony accordingly is formal irony (or what Friedrich Schlegel called irony of irony). Its function is to preserve the limited, contingent work through the destruction of its form, that is, the formal closure and illusion of first-order observation. Since, as we have seen, second-order observation cannot dispense with first-order observation, the destruction of illusion also preserves, even potentiates illusion—a paradox particularly evident in drama but also brilliantly explored for instance by Italo Calvino in *If on a Winter's Night a Traveller*. The destruction of illusion, that is, the ironic dissolution of the limited form, opens up the work to allow a momentary glimpse of the other side of form inside of form. Formal irony is for Schlegel the objective instance of self-criticism that expresses the spirit of art, and not the empty subjective will and arbitrary play of the artist as Hegel was to charge. It is the free spirit that hovers over the work and objectivizes it by manifesting the relation of the contingent work to the idea of art. In Benjamin's words, "The ironizing of the form of presentation is as it were the storm which raises the veil before the transcendental order of art to reveal it and in it the immediate existence of the work as a mystery."[9]

The mystery of the paradox of form returns in Luhmann's concept of world art. All distinctions make the unmarked world invisible and leave us with the paradox of the invisible unity of the distinction. Luhmann suggests that the work of art symbolizes the world by inducing the observer to see a form, behind which lies hidden the unmarked state of the imaginary world (comparable to the imaginary numbers of mathematics). Like religion, art is fascinated by what escapes it. The vision of God is the crossing of the boundary, the re-entry of form into the form, the making visible of the unobservable in the observable—as incarnation, as sacrament, as miracle. Art, however, is not religion. It seeks to explain the world from within. Art must remain within form, and it is only within form that the re-entry of form can be made visible.[10]

How does literature do this, how does it manifest its own self-limiting rationality? Romantic irony is the theory of re-entry and at the same time a theory of modern literature. For the German romantics, *the* form of modern literature is the novel, since it mediates and dissolves all forms, all genres and styles, in its form. The romantic theory of the novel thus comprehends the theory of modern literature. The anticanonical form of the novel exemplifies the break with the closed and completed forms of traditional literature, which break brings with it a reformulation of the code of literature. In the traditional genres, form served the expression of beauty; now the open, irregular form of the novel, which permits the highest degree of the self-limitation and expansion of reflection, fuses poetry and criticism and serves the expression of the idea of art.[11] All the recent celebrations of postmodernist metafictions have their origin in the romantics' "invention" of literature, which is its own theory. But if postmodernist metafiction points back to romantic irony, romantic irony in turn draws its inspiration from earlier literary practice. Schlegel's cardinal witnesses are Aristophanes and Cervantes, his models of the two "open" genres of comedy and the novel in the two very different media of theatrical performance and print.

The Self-Reference of Drama

Self-reference is the product of second-order observation, which distinguishes between self- and external reference. This can take the form of external or self-observation. Insofar as the distinction is external, the play remains within the "illusion" of first-order observation, while yet being predicated on the higher observational power of the audience. Drama is thus inherently self-referential since the other side of its form is the invisible audience to which it plays. The logical endpoint of this two-sidedness of form would be the reversal that turns the audience into the observed and the actors into the observers. This extreme example demonstrates that the theater is the frame of drama's two-sided form, the "wooden O" that contains the drama's world in the world.

Even if drama's self-reference remains invisible like the audience on the level of first-order representation, the invisible audience is made "visible" on stage through the blindness of the participants in the dramatic action. In fact blindness is the indispensable and inexhaustible theme of drama, and we can say that the making visible of blindness is the implicit (latent) and thus ever-given self-reference of the drama. The apparently closed dramatic illusion calls forth dramatic irony, which rests on the difference in information between participants in the dramatic action and between them and the audience. As Schwanitz reminds us, the fundamental ingredients of plot are intrigues, simulations, deceptions, disguises, confidences true and false, the deceiver deceived and so on.[12] The figure who is deceived or "framed" cannot see the frame in which he is entrapped, but of course the audience is intended to see his blindness. From the oracle to the whispered aside, dramatic irony exploits the difference between those inside and those outside the dramatic action. Drama foregrounds the blindness of observation and the observation of blindness. The invisible re-entry of form in first-order representation thus appears in the play as the blind spot of all observation. The visibility of this blindness is heightened by all the forms of "invisible" observation on stage (spying, eavesdropping, etc.), which duplicate the invisible audience.

The visible re-entry of form into second-order representation raises the implicit self-reference of drama to explicit self-reflection. The play within the play presents itself as play within the theater and as theater within the world. The knowledge that all the world's a stage is the frame of frames, which can only be made visible by the frame within the frame. The play within the play demonstrates before our eyes the paradox of form: that less is both less and more.

The paradox of the second re-entry of form is this: the play, which is contained and framed within the play, *contains and frames at the same time the play which contains it.* The form *in* the form is the form of the (containing) form. (And this applies also to the film in the film, the painting in the painting, the book in the book, the poem in the poem, the story in the story.) The invisible boundary, that is, the invisible distinction that separates audience and play but also play and world, is made visible through the re-entry of form, that is, through the self-limitation of form. The play within the play can manifest within itself the comedy (*Midsummer Night's Dream*) or the mystery (*Six Characters in Search of an Author*) of the crossing of the boundary between the inside and the outside of form, between illusion and reality, play and world. This crossing of the boundary can transform the paradox of form into the paradox of reality: does the world frame the play, or does the play frame the world? Is the play but such stuff as dreams are made on, or is life itself a dream? What is involved here is the implication for the observer of a self-implicating form. Let me illustrate this by reference to Corneille's play *The Liar*. On the first level we observe the comedy of an ever-more-improbable tissue of lies; on a second level we can observe that the play itself—and thus drama—is presenting itself as a lie, as a fiction; and on a third level we might ask what this presentation of the world in the world implies for the world of the observer.

What we might call the paradox of reality can only appear, however, if it is put into form, which takes us back to the play *The Liar*. And form, as we know, cannot exist without an observer, that is, without a distinction. And this is the answer to all the paradoxical theories that declare contemporary society to be a society of the spectacle, a world of simulation or pastiche in which reality has become indistinguishable from illusion. The answer is the simple question, where is the observer in this theory?

The Self-Reference of the Novel

For all its precursors, brought to our attention by Mikhail Bakhtin, the novel, unlike the drama, is a specifically modern form, indeed for the German Romantics *the* form of modern literature. If drama's self-reference lives from the presence of theatrical performance, the novel's self-reference is activated by the distance created by its medium: the printed word. But like the drama, the visible re-entry of form—the book into the book—serves to thematize the difference between the two sides of form—fiction and reality, world and World—within the form. As a specifically modern form, the novel is constituted by the dual process of the self-referential internalization and the externalization of second-order observation. By internalization I understand the (external) recognition of the social autonomy of the novel (the constitution of literature as a separate social system of communication). The social autonomy of the novel is gained by its differentiation from all other forms of narrative observation (history, chronicle, biography, memoir, narratives of exploration, ethnographic accounts, etc.) or documentation of the world (diaries, letters, logbooks, etc.). These are all forms of first-order observation that the novel takes over and incorporates into its own form by "doublevoicing" them, to adopt Bakhtin's formulation. That is to say, the social autonomy of literature is the outcome of the process of the differentiation, that is fictionalization of the novel in the seventeenth and eighteenth centuries.[13] Its inner literary complement is the externalization of second-order observation in the self-reflection of the novel as form. Through self-reflection—the visible re-entry of form—the novel demonstrates its autonomy in relation to the world.

World autonomy signifies the break between the novel and the world. As Schwanitz puts it, narration loses its innocence.[14] That is to say, narration has become conscious of its form, and it is this that distinguishes it from traditional narrative forms in which, in Bernhard Giesen's words, the "relation between narrator and listener, the relation between the components of the narrative and the lifeworld of the listeners, are constructed according to the same principles."[15] The homology between narratives and lifeworld presupposes of course the distinction between the two but leaves the problem of narrative form invisible (latent). This problem can emerge only when the distinction between the narrative and the world is drawn within the narrative. And here of course the great example is *Don Quixote*. From this point on the novel can distinguish between self-reference and external reference and can use the difference between first- and second-order observation for purposes of self-reflection.

If the novelist orients himself to reality (external reference), then we have the "realistic" novel, which distinguishes itself from all "unrealistic" forms of narrative. A whole series of conventions are elaborated in the eighteenth century to authenticate the veracity of the narrative, such as narrative or editorial frames that refer to collections of documents, reports, letters, and so on—a method of appeal to authentic documents, which is already parodied in *Don Quixote*. Cervantes patents, as it were, the other possibility of the distinction between external and self-reference by making self-reference the key to the realism of the novel; that is to say, the novel authenticates itself by foregrounding the difference between fiction and reality within the fiction. In other words, in contrast to traditional narratives, the question of realism (external reference) is defined in terms of the unity of the difference between self- and external reference.

If we analyze the two types of novel in terms of the observer, that is, the reader, we can say that the difference between the realistic line and the self-referential line of the novel lies in the difference between the implicit and the explicit reader. The self-referential novel makes explicit what remains implicit in the realistic novel—the difference between first- and second-order observation. In other words, the realistic novel constructs its implicit (or ideal) reader as the reader who can observe the difference between first- and second-order observation, the difference between the manifest and the latent. The implicit reader is called upon to observe not only the different perspectives of the characters and their respective blind spots but above all their self-blindness. The observation of action is deepened and complicated by the question of motivation. In other words, the novel discovers the social and personal "unconscious" of its characters and produces by means of this difference its ideal reader who comes to understand that the world is not what it appears to be.

The self-referential novel observes itself observing the world and discovers thereby the latency, the "unconscious" of its own form, which it makes manifest through the re-entry of form, through the demonstration, that is, of the limitation of its form. A central example is Tristam Shandy's attempt to recount his life in all its detail. What Laurence Sterne demonstrates is the impossibility of eliminating the difference between the two sides of form. His project is the equivalent of a map of London as large as London. Thus what he proves negatively is the necessity of form. The map of London is not London, but it has a much higher observational power than the attempt to see London from the ground. To repeat: the form within the form frames the enclosing form.

The realistic novel is directed, as Schwanitz notes, to the exploration of the conditions of the possibility of experience.[16] The particular means of observation it develops to this end is the alternating observation of a figure from outside and from inside. This we might call the paradox of realism. The self-referential novel uses the paradox of form—self-reflection from outside and from inside the form—to explore the conditions of the possibility of narration.

The Code of Literature

For Luhmann the code of art is to be sought in the self-referential use of the difference between self-reference and external reference. The unity of this difference exists in two forms. It is first of all the two sides of form, that is, the severed unity of the world, and secondly its correlative, which is the re-entry of the form into the form, that is, the self-reference that distinguishes between self-reference and external reference. "The world as the correlative of self-reference withdraws behind the distinction between system and environment: like observation itself it remains inaccessible to observation."[17]

The difference between self- and external reference reformulates the traditional distinction between form and content. The re-entry of the form into the form serves both the production of this difference and its theoretical observation. Re-entry allows us to define the relation between literature and the world (the "truth," the "rationality" of literature) as the unity of the difference between literature and the world within the forms of literature. Where we stress the *unity*, the internal difference remains invisible (latent) and form appears as content. The observer is the invisible observer of first-order observation. Where we stress the *difference*, form appears as the form of the content and we observe second-order self-observation.

This difference is not a difference of quality but an evolutionary difference of kind, the difference, in Luhmann's terms, between object art and world art. From an evolutionary perspective we can understand it as the difference between the hierarchically stratified forms of premodern societies and the contingent, experimental, and open forms of literature in functionally differentiated societies, the difference, that is, between a premodern hierarchy of genres and the modern system of literature. "Objective" or first-order coding entails the priority of generic models, which are monologic, in Bakhtin's terms, in that they present a unity of form and content underpinned by the unity of a worldview. "Subjective" or second-order observation entails the priority of direct or indirect self-referentiality, the breakdown of generic models and the emergency of the *sui generis* work. (We should not forget that the modern writer has the choice between "naïve" and "sentimental" forms. The option for "naïveté," however, cannot be a naïve choice.)

Such a general formal-historical typology is of course only a first approximation, which remains to be developed and modified. The question it poses to both premodern and modern literature is the question of self-reference. Where self-reference remains latent, where the form lacks self-consciousness, this absence can be rectified by an external observer. This is the function of parody, which, as the name indicates, is song alongside the song. Parody is the corrective of first-order observation: the critical observation of the organizing generic forms of literary observation from the heroic epic to the soap opera. Parody duplicates with a difference, which it demonstrates by highlighting with comic-critical intent the incongruity between style and subject matter, form and content, in the object parodied. In short, parody is the supplement, the external "self-reference" of first-order observation.[18]

The paradox of form points back to the observer. Let me conclude with a reference to a visualization of the paradox of observation by that master of the paradoxes of self-reference—Magritte. The painting, appropriately titled *The Human Condition I*, is elucidated by Magritte as follows:

I placed in front of a window, seen from a room, a painting representing exactly that part of the landscape which was hidden from view by the painting. Therefore, the tree represented in the painting hid from view the tree situated behind it, outside the room. It existed for the spectator, as it were, simultaneously in his mind, as both inside the room in the painting, and outside in the real landscape. Which is how we see the world: we see it as being outside ourselves, even though it is only a mental representation of it that we experience inside ourselves.[19]

What Magritte's painting shows is that the spectator is only brought to see how he sees the world through the re-entry of the form into the form: the painting of the painting of the tree, the window as the visible/invisible boundary between inside and outside, which in turn is visibly/invisibly replaced by the painting before the window. Magritte's painting neatly demonstrates that the difference between inside and outside, self-reference and external reference, is a difference within the form. This difference is both invisible—the canvas in front of the window is not distinct from the landscape behind it—and visible—the canvas rests on an easel, one edge separates it from the landscape. The juxtaposition of first- and second-order observation here constitutes the difference that frames this unframed canvas. Moreover, as we observed with the play within the play, the painting within the painting frames the painting that contains it. This frame, constituted by difference, is the code of art.

NIKLAS LUHMANN | **SIGN AS FORM**

I

That human orientation to the world uses signs, indeed is bound up with signs, has been known and discussed since antiquity. The concept of the sign was first and foremost supported by a certain familiarity: that signs abound in the world was considered common sense. The word "sign" thus designated something that realizes a certain mode of being—an essence—in Being. More precisely, signs serve to make intelligible what is not in itself observable. This is reflected, for example, in the medical usage of the terms *sēmeion* and *signum*.[1] Therefore, signs could be distinguished from other sorts of things and investigated in their specificity.[2] Rhetoric, for instance, distinguished between *verba* (words) and *res* (things).[3] This consequently led to a sub-ontology of sign-using beings and, in this context, to an ontology of language. Both knowing names and giving names was thought to require a certain artistry—in particular, a knowledge of the nature of things. And the same holds for writing.[4]

Only during the last hundred years or so has the theory of signs apparently acquired universality and therefore radicalness. The names and theoretical programs are well known: Peirce and "semiotics"; Saussure and *sémiologie*. At first glance, this universalization and radicalization seem to imply that there are only signs and nothing else. Bernhard Giesen calls this the de-reification of the social.[5] Systems emerge through repetitive use of signs (Peirce's "iteration") as a processing of "tokens." For them reality is given in no other way than in the facticity of this processing. This undermines a tacit premise of the aforementioned ontology, namely, the assumption that for every existing thing there always exists something of another kind. Otherwise there could not be differences between different beings. More cautious radicalizations claim only that there is no one-to-one relationship between signs and things, but rather a sort of overall correspondence between sign usage and the surrounding world. Another kind of radicalization insists that the world is only known and familiar to us through signs. Such a position amounts to a constructivist epistemology. But does that necessarily mean—as is often inferred—signs without referents? And would such signs without referents still be signs, or would that universalization and radicalization dissolve the meaning of the very concept of sign? And would that not inevitably lead to the paradox (and why not?) that the word "sign" and/or the concept of the sign is, in turn, a sign without reference and, consequently, a sign expressing that there are no signs?[6]

That may be, and if so, this would mark not the end but the beginning of a theory of signs. A theory of signs would then have to deal with its own paradox, unfold its paradox, and replace that paradox with a distinction. Only the unity of this distinction, but not what is distinguished, would then have a paradoxical status that blocks observation.

In this regard, Josef Simon recently proposed a noteworthy hypothesis,[7] which takes as a fact that the sign cannot "say" anything about the signified. The signifying function runs, as it were, on empty. What might be called meaning fulfillment in Husserlian terms is replaced by a distinction. Either the sign needs an interpretation, thus implying a need for further signs to which the same applies; or the sign is immediately comprehensible, or at least understandable to the extent that, in acting, one can grasp the sign and risk using it. However, the immediacy of understanding can only be achieved within a particular situation, hence at a particular moment in time. Doubts, further queries, and calls for interpretation cannot be prevented from arising in the next moment.

Once again, the paradox of the nonsignifying sign is unfolded by means of a distinction. Here, however, the signified is completely excluded from consideration as something unreachable, and it becomes questionable what the meaning of a sign might be when we realize that it does not recognizably signify anything. Having its precursors in the philosophy of life, in the immediacy of the self-to-self relationship of reflection theories, and in Heidegger's analytic of *Dasein* (being-there), the discourse of immediacy, in truth, only designates the collapse of certain distinctions, be it that of subject/object or that of sign/signified. This discourse of immediacy thus presupposes that one has to begin with those distinctions, which are then not used in the immediacy of the operation.

However, the function and the exact whereabouts of what is signified by the sign remain unclear. Consequently the meaning of the sign concept related to this difference remains unclear as well.

For that reason, I begin with this difference. Common linguistic usage is content with a concept that characterizes this difference as a *relation* between the sign (or the signifier) and the signified.[8] If one accepts the traditional understanding of the concept of relation, such a conceptualization suggests taking the relation as secondary with respect to its relational elements. However, semiology has the exact opposite conceptual understanding of a sign since it is the sign that constitutes the difference between signifier and signified in the first place. To mark this turn (of course, not to be understood in a "holistic" sense), I shall replace the term *relation* by the term *distinction*. In that manner, it becomes clearer that the distinction takes the place of the paradox and that the distinction, in unfolding the paradox, renders the paradox invisible at the same time. I shall designate the unity of the distinction that takes over this function of unfolding the paradox with the term "form" so that we are able to *distinguish* the unity from the identity of what is distinguished in each instance. In this regard, I subject the concept of sign (and hence semiology) to an analysis of form.

Before doing so, I need to state that I take system reference as a given.[9] Signs *per se* do not exist; rather they exist only as forms within the operations of a system that uses them. Saussure had already assumed that it is not the physical materiality of the sign—for example, sound or image—that fulfills the function of signification, but exclusively the sensory impressions triggered by it. Though this can only be understood psychologically, it nevertheless hints at the fact that the signifier and, with it, the distinction between signifier and signified (that is, the sign) are structures of a system capable of operations. Thus they do not occur within the environment of the system. The environment of the system does not contain signs at all; it does not even contain distinctions, nor does it contain any kind of information. The environment is as it is, and nothing more can be said about it.[10] In every instance therefore, each sign is always a system-internal structure. This does not restrict us to psychic systems as our starting point, for we might also choose to begin with social systems. In such a case, we are concerned with communicative structures and mainly, although not exclusively, with language. The decisive factor is that semiotics can be—indeed, must be—related to a theory of operationally closed systems since semiotics *deals with distinctions*.

II

As in the case of Spencer-Brown's calculus of form,[11] the concept of form will here be related to a presupposed distinction. Provided the distinction is distinguished from what is distinguished by the distinction,[12] the distinction itself is the form. Now we can employ the notion of sign to designate the form of a particular distinction, namely the form of the signifier/signified distinction. In order to indicate the two-sidedness of the sign, we therefore require three terms: signifier, signified, and sign. It is quite possible to capture the basic idea of this conceptualization with the following definition: A sign is the difference between the sign and the signified (or likewise: a system is the difference between system and environment; a distinction is the distinction between distinction and indication). Such a definition would point to the fundamental paradox of the "re-entry" (Spencer-Brown) of a distinction into itself. However, since it would be difficult to know which level of self-implication is intended, this reduction would be confusing for the normal application of any theory dependent on the unfolding of this paradox. Every unreflected attempt at reducing the threefold distinction to a twofold one (i.e., sign and the signified) would make the problem of such a distinction—and hence, semiotics—disappear.[13] Each user of the sign form—and of every other kind of distinction—must indicate to which of the two sides of the distinction he is referring.[14] One must choose one side or the other. According to the axiomatic framework, the user of the sign form cannot designate both sides at the same time since this would amount to the paradoxical nondistinctiveness of what is distinguished. Certainly, the user could designate a distinction, such as the one between signifier and signified; however this can only be done within the context of a second distinction, such as the sign/system distinction to which, then, the same would apply.

Saussure had already arrived at a similar conclusion: According to his theory, *signifiant* and *signifié* are the necessary components of a linguistic unit, that is, of a sign. They differ not with respect to their nature or essence but only as components of just such a distinction. One cannot exist without the other. Subsequent discussions did not always pay appropriate attention to this—perhaps insufficiently elaborated—insight and as a result formed, for instance, the idea of a sign without a referent (which in English is often termed a "token"). Nevertheless, an important difference is overlooked. There is indeed no referent for the sign as form; which is to say: one can either make use of the signifier/signified distinction or not. There is no "external" point of reference that would force one to select either option; neither is there any truth criterion for choosing a first distinction as a starting point.[15] That is why a theory of language constructed as semiotics must relinquish the idea of language's external referent. (Here we notice parallels to a theory of operationally closed, autopoietic systems.) As has been recognized since Saussure, all differences are differences inherent in language. However, this does not free us from the necessity of retaining the signifier/signified distinction. There cannot be a signifier without a signified. The distinction itself, and thus the sign, would collapse if it had only one side. Form is always two-sided.

Also excluded is an understanding of language (or other systems structurally and operationally based on signs) as a manipulation of signifiers. In other words, language is always "world language," that is, language about the world. Language only maintains its sign-using form as long as it relates signifier to signified. Language feeds on this distinction, thereby always constructing something other than itself. This does not imply, though, that language can operationally impact the outside world or perform functions of representation or adaptation. It is at this point that Peirce's critique becomes pertinent. Just as language has no operational impact on nonlinguistic realities, neither does the external world have any impact on language. Yet language does not speak only about itself.

Hence, the "problem of reference" is poorly framed. Systems theory may be of some help at this point. An operationally closed, language-using system cannot interface with the environment on the level of its operations. In this respect, reference does not exist. But this impossibility is internally compensated by the system itself in that in observing its operations it can—in fact, must—distinguish between self-reference and external reference. Indeed, the semiotic structure of language explains how this happens, namely, by means of language's use of the other side of the sign form (i.e., the signified) as an external reference. In other words, by using signs exclusively, language very successfully solves the problem of how a system can constantly force itself to operationally couple and simultaneously process both self-reference and external reference.

III

Let us once again clarify the consequences of the theory of the two-sided form and its application to the case in point, that is, signs. If the sign itself is the distinction between signifier and signified, then it is neither signifier nor signified, but something else. Understanding this third position is consequently one of the most difficult problems of semiotics. This problem is more clearly stated in the Peirce tradition than in the Saussurian since Peirce, referring to the third, did speak of an "interpretant"—a concept that includes the possibility of the interpretant's being, in turn, a sign. The interpretant identifies itself with the signifying function of the signifier by interpreting this function as a relation (or as a representation) of signifier to signified. The interpretant thus is a representation of representation.[16] For Peirce, the interpretant is both the meaning of the sign and, therefore, in a dynamic context, its pragmatic reference. Yet this choice of words also suggests interpretations (which would be interpretations of the sign "interpretant") that go beyond Peirce's own reading—for instance, interpretations leaning toward "linguistic community."[17] Whether those difficulties can be overcome by means of a thorough exegesis of Peirce is doubtful, but that task we leave to the philosophy expert. We start our project already questioning: In what precise sense can one actually speak of "thirdness"?

In any event, one has to start with the assumption that a distinction (such as that between signifier and signified) cannot be used as both a difference and a unity because doing so would result in one's being obstructed by a paradox. The use of distinctions (including the one between signifier and signified) therefore serves to dissolve this paradox. Specifically, one resolves it by recognizing what is distinguished as distinguished and by forgoing the possibility of thematizing the distinction's unity within the same operation.[18] Consequently, the distinction itself serves as a blind spot, as it were, that determines what one can or cannot see by means of the distinction.[19]

In order to replace the expression "interpretant" completely (rather than simply referring to the form as the unity of a distinction), I shall call *observing* the use of a distinction to signify one (and not the other) side.[20] This opens the possibility of *signifying* the user of a distinction as an observer. Since this leaves completely open the question of which types of operations are materialized by observations, one needs to *distinguish* carefully between operation and observation. Operations can be neurophysiological, conscious, or communicative; we could even include living cells if we wanted to consider them observers. Consequently, speaking of something as an observer presupposes the determination of a reference to a system. This then also implies the dissolution of the concept of subject that, being based on consciousness, could only recognize observers with a world status. This dissolution, in turn, "deconstructs" the subject/object distinction—including the assumption that all observers, as subjects, observe a collectively pre-given world and that they in their intersubjective relations merely adjust their various "lenses" (logic, empirical methods, theories, and so forth) from the perspective of the truth—or untruth—of the results.

This postulated objectivity and its corresponding "intersubjectivity," which together signal a relationship between subjects capable of error-free cognition, is now replaced by the possibility of observing observers with regard to what they can or cannot designate with each applied distinction. Such a "second-order observation" is equivalent to any other observation. It uses distinctions as well, namely the distinction made between the various observers to be observed, thereby distinguishing between the distinctions those observers employ. Once again, signs must be available for this second-order observation, that is, signs that are specialized in distinguishing distinctions. Neither any exception to the general conditions of observation nor any "higher" or "better" knowledge is claimed for second-order observation. Unlike the deparadoxifying theory of types and its attendant multileveled analysis, this theory assumes no hierarchy. However, one has to take into consideration the fact that second-order observation puts demands on logic that the usual binary logic can hardly meet.

IV

Having briefly considered the epistemological scope of the context within which our conceptual sign analysis (based on a theory of difference) is placed, I now return to semiotics in the narrow sense by recalling that the sign itself has no corresponding referent. This can also be reformulated as the requirement for *isolation*. Isolation can only be achieved by *arbitrariness* in the coding of signs: Saussure's famous "l'arbitraire du signe" (arbitrariness of signs). In a less differentiated sense, one could certainly speak of "natural signs" in terms of extrapolations from perceptions. One perceives which way the wind blows; position and movement serve as signs for something not yet visible. Similarly, cloud formations indicate to the observer an approaching thunderstorm. The technique of reading signs could not have evolved without such connections surmised from the signified. The case of interest to us here, however, is based on further differentiating the form, that is, on isolating the distinction between signifier and signified while uncoupling other possible references of meaning in the world of perception. Mind you, this case does not focus on the particular qualities of those components used for signification, such as words in the context of language. Rather, isolation pertains to the form of the two-sided sign—the difference between the signifier and the signified. Only thus is it possible to understand the explosion of meanings ignited by the evolution of language. With language, not only does a new type of object with the function of signification (i.e., words) enter the world, but the entire world is rearranged by a pool of strictly isolated sign forms (which are then to be recombined) that force the user always to bear in mind both signifier and signified. This is the only way in which the iterability of using signs can be ensured.

As with any distinction, the form of the sign only contains itself. Spencer-Brown has suggested, "Distinction is perfect continence."[21] This might just mean that something specifically signified excludes everything else and, in this sense, is something-in-the-world. In the particular case of the sign form, the same logic of containment means that the signifier designates only the signified and nothing else, just as the signified is designated exclusively by the signifier.[22] In this regard, the form is complete and closed and, precisely because of this, isolated. The possibility of a multiplicity of signs, and correspondingly of operations finding a way from one sign to the other, is by no means excluded. However, this then must be corroborated at the level of distinguishing distinctions and should not be confounded with the sign function that rests on isolation.

We are apt to think of arbitrariness and tradition as being opposing terms. The reverse is true. Once again, a more precise analysis of the distinguishing structure of signs can help clarify this point. The necessity of an arbitrary relationship only pertains to the relation between signifier and signified; here the required isolation does not come about without arbitrariness. However, the sign itself must be capable of reiterative use as a discernible unit and thus capable of being remembered. The sign cannot be invented anew with each and every use.[23] The arbitrariness of the relationship between signifier and signified specifically precludes this option. Tradition, or analogy as Saussurian structuralism has it, then comes into play where relations between signs are concerned.

And so does redundancy. If signs are to be combined with signs, for the purpose of communication and thought, for instance, then expectations have to be directed and the possibilities of further connections limited. The subsequent sign must not be predetermined, nor should it be too surprising. Each sign must, therefore, not only function as an entity by itself but also provide redundant information. In other words, it has to inform what might ensue. With its limited information value, the next sign then specifies and confirms what had been expected.[24] However, the required redundancy can only be established if signs have the capacity for reciprocal limitation, that is, if from the outset they are given a high degree of surplus meaning, *ergo* if they are initially set up arbitrarily and not restricted by nature.

This analysis recognizes that arbitrariness as such does not exist.[25] If we conceive of arbitrariness as the determination of events (decisions) by decisions free of any structures or context, arbitrariness would equal entropy. Accordingly, all that can exist is what is denoted by concepts such as isolation, uncoupling, nondifference; this in turn generates a transitory open space for "arbitrariness" that immediately seeks other limitations. It is precisely in this sense that the arbitrary relationship between signifier and signified is possible only in the form of signs—an arbitrary relationship subjected, in turn, to limitations of use. The ultimately paradoxical form of the unity of what is different explains how and why it comes to, in fact must come to, a combination of arbitrariness and redundancy stabilized by tradition.

V

Distinctions are built asymmetrically. At least, in the very moment of their use (and they are only available when used), one side of the distinction and, consequently, not the other is designated. In linguistics, the term "marking" indicates that one (and not the other) side of a distinction should be dealt with as a problem (and should not be taken for granted).[26] Following Spencer-Brown, we can call the marked side of a form its "inside." This asymmetry can also be found in the relationship between signifier and signified. Connecting operations take place on the inside of the form and employ markings; one only speaks with words, and this remains as long as the underlying meaning of a speech act remains unspoken. Only language theories, by conceptualizing signs as forms, establish themselves as a special case within semiotics.[27]

Consequently, the sign is not a reversible distinction. If one wishes to designate—not merely to carry along—the other side (i.e., the outside of the form), one has to employ yet another sign, which in the most general sense would be the sign "the signified." This is even true when one seeks to indicate something immaterial such as values. Values are typically found on the unmarked side of the form. They are presupposed, taken for granted; and normally one is concerned only with the valued behavior, rather than with the value itself. When, however, values are designated, the question naturally arises of whether they will be accepted or rejected, for example, in order to identify who is taking what side. That is why the analysis of values is always an analysis of the interest groups involved. The semiotics of the sign form destroys all *a priori* value assumptions by constantly bringing the analysis back to the way we deal with signs. The same pertains to "reason."[28]

Such reflections might suggest "postmodern" or "deconstructive" consequences. Even if those consequences do not give us pause, one still has to keep in mind the specific architecture of the sign (i.e., its form) and must stress the precision of the analysis. We ought not follow the path that Roland Barthes pursued with exemplary rigor, namely, that of depriving the signifier of its reference and, as a result, viewing the signifier as the sign itself—that is, a sign without a referent.[29] To avoid this impasse, one has to keep in mind the difference between the sign as a distinguishing form and the sign's distinguished components, that is, signifier and signified.[30] This is important because it is not possible (or it is possible only when a blocking paradox is accepted as the price) to designate within one operation both the unity of the form and the components distinguished by the form. The paradox that would then result, to borrow from Ranulph Glanville, could be termed "the same is different"[31] since signifying always implies making use of signs—designating one side (and not the other one) within the context of a (perfectly contingent) distinction. To designate the sign as the unity (form) of a distinction thus calls for a contextual change requiring different distinctions, different operations, and above all time.

One consequence of the analysis of sign as form is that semiotics becomes able to designate itself—to apply its own concept of the sign to itself, reflexively. However, for that to occur, semiotics has to adopt a difference-theoretical stance. In that case, semiotics can no longer consider itself a science of particular objects called signs. Along these lines, Dean and Juliet MacCannell call for a second-order semiotics, that is, a self-inclusive semiotics of difference.[32] "Signs are disunities;"[33] every theoretical quest for an ultimate, all-encompassing unity can now, in turn, be designated as a "desire to return to a state of nature."[34]

VI

"Sign" is itself a sign. Thus semiotics, since it contains itself as one of its own objects, is an autologically constituted science. Furthermore, a sign is not a thing, but rather a distinction. Accordingly, being autologically founded amounts to being based on a difference. This has consequences for determining which conceptualization of the world at large semiotics might reasonably find acceptable. The world can no longer be considered the entirety of signifying and signified objects and the relations between them. Concepts of substance, as well as concepts of relation, are dissolved by the interpretation of the sign as form, or they occur—to the extent that they are still used—only as signs. Hence, it is possible to *signify* the world only as that which is bifurcated by the form of the sign. That is to say, one must presuppose the "unmarked state" in order to begin positing signs as a specific way of making distinctions. This notion of the "unmarked state" again refers to Spencer-Brown's calculus of form.

The unfathomable is difficult to grasp. As a form, each signifier already has an outside, namely, what it signifies. Now we have to consider yet another outside, namely, the outside of the difference between signifier and signified, that is, the outside of the unity of this difference, the outside of the sign. This would then be the world. Perhaps the theoretical need for closure forces us to follow this line of reasoning. Nonetheless, we must also keep in mind that, for a radical approach based on a theory of difference, the world concept cannot possibly be otherwise positioned. The "unmarked state" refers to a placeholder for what cannot be distinguished but can only be brought into a form by making a distinction. As a result, there are always further distinctions and other possible forms and significations.[35]

It does not follow from this that a sign-using, communicating system must—or even could—operate without boundaries.[36] The world concept resolves this problem. For it is only the world that cannot be conceived in terms of boundaries or as a form; neither can it be conceptualized in tandem with anything on the other side of the boundary.[37] Systems, by the very fact of their use of operations, cannot help but establish boundaries. Nonetheless, they reproduce themselves, organize themselves, and generate their own structures and boundaries by their own operations. Consequently, an observer outside the system could not detect the location of the system's boundaries, even though their existence is unquestionable. Otherwise, no system could materialize. To state my position from another angle, any system constitutes itself as a form, as a boundary, as an asymmetrical difference between system and environment. And if a system has at its disposal the appropriate capacities for reflection, it can use this form (which, as we have shown, is the system itself) in order to signify, observe, and describe itself as different from the environment. But how does such a system handle the world, if the system cannot signify the world (or can signify it only paradoxically as the unsignifiable)?

This is the moment to take advantage of the circumstance created by freeing the concept of meaning from the concept of the referent to which it was formerly tied. As a form, a sign has no referent and thus, one could infer, no meaning. But for some time now the concept of meaning has been uncoupled from the concept of the referent and redefined within phenomenology. We have only to return to phenomenology's description of the phenomenon of meaning and to expand on it.[38]

Phenomenology describes meaning as a surplus of references, that is, as references beyond what is intended at any given moment.[39] These references cannot be brought to a definitive closure. Instead they lead into horizons for which the world serves as the final horizon. The world is thus to be conceived of as the counterpart of meaning, as a world illusion, if you will, that offers a closure—a closure that in actuality is never achieved—to the ceaseless continuation of all meaning references, including self-referential ones. In the language of form analysis, meaning can also be defined as the unity of the differences between actuality and potentiality in experiencing or communicating, thus as the form of this distinction. This is just another version of what phenomenology describes. However, the concept of the two-sided form better expresses the fact that the inside of the form—namely, the meaning actualized in each instance—makes sense only with respect to the possibility of actualizing other possibilities, and the fact that this presupposes the existence of dynamic systems composed of operations (events). Therefore, meaning is form as boundary that, while always being observed alongside, can never be operationally transgressed since every meaning-actualizing operation remains on the inside of the form. We thus rediscover a fact we had previously encountered in our analysis of the sign form: operations remain on the inside of the form. Transgression of the boundaries would signal the disappearance of the operations (Spencer-Brown's "law of cancellation"). Operations manipulate signs; however, this can only occur when the form functions as a form, which is to say, when the form provides a second side.

In the end, this other side always remains an "unmarked state." There is no other way that the sign-using operation can take place in the world or confine itself within the world. At the same time, each particular sign can be—in fact must be—operationally understood as an instruction for crossing the boundary, and this is only possible in reference to something specific or to something that can be specified, which is to say, with respect to something signifiable. The structure of the signifier/signified distinction meets these requirements since the structure lets the signified—from the viewpoint of the signifier—appear as if it were specified, even though the signified is not *in fact* specified. This specified/unspecified ambiguity of the signified (or to put it phenomenologically, the signified as identifiable only in the world, only by virtue of a surplus of reference, only through further selections) threw semiotics into considerable confusion and eventually lead semiotics to repudiate reference altogether. Here, too, further distinctions might be of some help. Operations using the sign form always take place on the side of the signifier. However, to be able to mark this side, they require another side. This other side is always present as an unmarked state. This way, the other side assures the simultaneity (simultaneous presence) of a world that remains unattainable (*inattingibilis*). At the same time, to be sure, the other side also provides the selection pool for the subsequent connecting operation. This operation needs to specify something from the domain of what is possible to be able to re-use that something semiotically as a signifier. Only in this manner can meaning be continually actualized without ever being exhausted. The sign form is one among many possibilities able to translate—and therefore, to unfold—the paradox of meaning (i.e., the nondeterminativeness of the specified) inherent in a distinction.

Finally, we can adduce yet another distinction, namely, that between *medium and form*,[40] in order to illuminate still other aspects of the phenomenon of meaning. This distinction presupposes the concept of an element and offers two different possibilities for coupling elements. Loosely coupled elements constitute a medium; tightly coupled elements constitute a form. Only by means of a form do actualizable meaning-structures differ from meaning as a general medium that merely provides the possibility for couplings to take place. Here again, forms that occur through the selection of tight couplings in the medium of meaning—which continue to exist and dissolve, be remembered and forgotten—can constitute forms only with respect to the other side of the form. Forms are forms only with respect to a medium that supplies an ample amount of momentarily, loosely coupled (i.e., unusable) elements for tight couplings. The distinction between medium and form is itself a form with an inside and an outside. This distinction belongs to the class of distinguished forms, such as sign (signifier/signified), meaning (actuality/potentiality), system (system/environment), or distinction/indication (Spencer-Brown's distinction), which all share the characteristic of containing themselves and which therefore can be directly observed only as a paradox.

What can be concluded from this for a theory of signs, that is, for semiotics? We have encountered a multiplicity of forms (distinctions) that coincide in the basic feature of self-implication and can therefore elucidate one another. The sign form is one such distinction. By comparing the sign form to other forms with the same paradoxical profile, its peculiarity can be ascertained. Systems operating in the medium of meaning, which are therefore required to observe and simultaneously process self-references and references to the environment, can be described as sign-processing systems. Accordingly, those systems are able to restrict their operations to the realm of the signifier since, thanks to the form of the sign, the signified is always carried along. Semiotics contributes the following insight to the theory of meaning as medium: The progression from one actuality to another in the medium of what is possible requires the sign form because only in that form can the selection of a tight coupling be accomplished. The sign ensures that, while a signification is selected, the signified is also sufficiently specified, even though the signified remains operationally inaccessible to the system. In other words, the form of the sign explains how one operation can be joined to the next, given that this occurs in an inaccessible world and in the medium of meaning, where an immense surplus of nonactualized possibilities is always available on the other side of its form.

These considerations dissolve historically evolved ideas of order in the systems domain of science. They are in part related to disciplines or areas of research such as semiotics or linguistics. In part, they make use of ideas from calculus that, despite their roots in mathematics, are hardly accepted there.[41] They are in part germane to theories (or "paradigms") that, while enjoying a certain interdisciplinary status as does systems theory, are nonetheless contested in the individual disciplines.[42] And they pertain in part to lesser-known special developments, such as the medium/form distinction, that stand in an unclear relationship to more common distinctions, such as entropy/negentropy or equilibrium/nonequilibrium (Prigogine). Thus there is a demand for interdisciplinary theoretical developments, which are, however, more hindered than furthered by the way in which forms relevant to such developments have been labeled and accommodated within the sciences thus far.

Incorporating the highly controversial relationships semiotics holds to both phenomenology and structuralism, Dean and Juliet MacCannell mentioned outlooks in their inquiries into second semiotics that can only be described as "postdisciplinary."[43] Nevertheless, their analysis remains restricted to the range of existing controversies. Still, to a much greater extent, there are unimagined possibilities of combination—encounters yet to occur—and there are connections too complex to materialize as controversy or as the type of patricidal frenzy seemingly indispensable for the progression of science. In this context, semiotics may have taken itself too seriously as a theory of linguistics or as an autonomous discipline, thereby pursuing its own isolation. Asking what it means to consider the sign as form could lead beyond that impasse.

VII

Our difference-theoretical clarifications of the fundamental concept of semiotics (i.e., the concept of the sign) now enable us to determine in which sense the notion "symbol" can still be used. Not in the least owing to the romantic movement, the concept of the symbol has been so widely circulated and broadened since the nineteenth century that it can hardly be distinguished from the concept of sign. Because of its theological tradition, the symbol concept seems suitable to imply depth of meaning, to signify more than does the matter-of-fact concept of the sign, applicable to any guidepost. Novalis once encouragingly affirmed that "symbols are mystifications."[44] In times of intellectual uncertainty, it may therefore be commendable (or at least it sounds good) to speak of symbols, symbolic forms, and so forth.[45]

If we consider the fact that a sign (insofar as it claims to be the unity of a difference, i.e., the nondistinctiveness of what is distinguished) conceals a paradox, then it is not completely unfounded to speak of mystification. This need not call for a respectful distance or a solemn mood. Returning to the original understanding of the word *symbolon*, one could rather surmise a very functional, if not pragmatic, meaning. *Symbolon* originally referred to a representation or evidence of a unity—above all that of a certain social standing earned by hospitality—with the help of two separate but matching pieces.[46] Thus, a *symbolon* represented a connection or correlation by virtue of these separated pieces. In the context of semiotics, this can only be construed as the distinctiveness of signifier and signified, a fact sufficiently covered (i.e., designated) by the concept of the sign. In cases where the sign itself designates its own function of unifying what is separated, one could speak of symbols. A symbol would then be the self-signification of a sign. Therefore, symbolic signs are not simply guideposts that point to something else. They are not just carriers of a signifying reference and, therefore, not only materializations of the signifier. Rather they also contain a hint of the function they fulfill, that is, a hint of the very unity-constituting meaning of the sign. Since symbolic signs degrade, as it were, the materialization of the signification to a mere component that does not yet constitute the "actual meaning" of the sign, only through symbolization is it possible to distinguish the sign itself from the signifier.

This specifically (and only then historically and practically) becomes an issue if what is signified by the signifier remains operationally inaccessible. Whatever a pointer refers to can never be reached, and it cannot be integrated as a link into the sequential chains of operations; as "meaning," for instance, it never becomes a "word." In language, a sentence can be followed only by a sentence and never by what is intended, meant, or designated by those sentences. To the extent that language reflects this and realizes that it functions despite (or precisely because of) this, language is handled as a symbolic medium.

It now becomes clear that the achievements of symbolization are evolutionary accomplishments; symbols are therefore historical outcomes. Adopting the concept of the symbol from everyday usage, the development of a religious capacity for differentiation certainly played a leading role in this historical process. In earlier societies (including the early period of ancient Egypt),[47] the category of the symbolic was not available. The sacred object was *itself* holy; the statues were *themselves* the gods. Only in conjunction with the development of mythical narratives could the idea of a second level, of a meaning proper, be consolidated—a meaning proper that the operationally accessible objects then merely "symbolize." But even then, Egyptian religiousness could not dissociate itself from its presymbolic roots and so retained those forms of devotion to the concrete object, upon which the Greeks were to comment with amazement. Even in the case of the interpretation of the Christian sacraments, as is well known, this problem of transition occurred once again. Only to the extent that priests became theologians and were in consequence confronted with problems of consistency in their observations (that is, in their cognitions *and* actions) did the sacred object become transformed into a sign and, because it becomes a sign for inaccessible meaning, a symbol.

In an exactly parallel manner, this process was repeated in nineteenth-century aesthetics. In permitting itself the liberty of granting a symbolic interpretation to the work of art, aesthetics had to allow *l'art pour l'art* as well, and generally had to yield to an extension of what was acceptable. To the extent that this occurred, it could then also be accepted that art forms symbolize something principally inaccessible, that is, an unobservable world.

If the two cases of religion and art suffice to suggest a generalization, they can confirm that signs provoke observers to observe observation. As for second-order semiotics, unity seen against the backdrop of difference becomes a problem for second-order observation as well. Only on this level can the transition from asking *what* to asking *how* be made. Only then does one escape being exclusively manipulated directly by signs, though one still is. Having made this transition, one asks not only *what* they designate but also *how* they signify what they signify. On this level, the concept of the symbol suspends further questioning. If one wishes to go beyond this state of affairs, one must ask why. However, if one does not intend to leave the realm of semiotics, this question could only lead to the paradox that any observation presupposes both the unity of a difference and the difference between observing and what is observed. Unless one wants to dissolve any and every thing, including oneself, into the "unmarked state" of the world,[48] the answer to the "why" can only be: because paradoxes have to be unfolded if observations are to be possible at all.

KARL EBERHARD SCHORR | ON THE ANALYSIS AND USE OF FORM IN LOGIC

The principle of the excluded middle (*tertium non datur*) was, in the traditional view, the most important conceptual feature of logic. Following Gödel's proof of "formally undecidable propositions,"[1] we can no longer take this feature as our starting point, and therefore must reexamine logic's controlling function in scientific inquiry. Conceived sociologically, such an undertaking naturally presupposes logic in terms of a science (of society). Moreover, it takes logic to be a theory that deals not only with its object but with itself. The reevaluation thus focuses on logic as a functional enterprise, for attending to functions is part of observing. And in view of the "Gödel catastrophe" there is good reason for this focus, since henceforth "truth" and "untruth" cannot be taken as universally applicable binary values. What happens, then, to the traditional definition of logic as a truth-validating device for scientific knowledge?

Guided by this question, my inquiry seeks to discover the function of logic after the Gödel catastrophe and by extension to examine the definition of logic's form. For this I base my inquiry on recent research on form;[2] yet my focus remains on the self-analysis of modern "formal logic," which prior to such research had already regarded the issue of form as a means of grounding itself. But what then is a form? And above all, how is the form of logic defined?

According to the traditional understanding of logic, forms are concepts that are able to regulate the truth and falsity of knowledge expressed in single or compound propositions because they are defined in reference to the notion of matter. Hence, in the classical view of logic, concepts are required to observe the principle of correspondence (*adaequatio*), however the notion of truth (nominalism/realism) may be defined. Accordingly, logical forms reflect an ontological understanding of truth whose underlying view of reality operates with the idea of correspondence or noncorrespondence; a third position does not exist. Yet this logical understanding of form became untenable with Gödel's proof. To critically observe the development of the analysis of form in modern logic, which eventually led to the disaster of undecidability, I shall begin with an approach to the problem of modern logic that provides a certain distance from this traditional view of reality. I see such an approach in Kant's discovery that the concept of form entails the distinction between form and matter—hence the potential of the Kantian form concept to overcome the ontological interpretation of this distinction (as I will discuss below in section I). How then do forms become logical forms? This question is taken up by the analysis of form through which formal logic attempts to set itself in motion. What is of interest here is not only that forms become concepts for dealing with truth, but also how they become those concepts. We are thus interested in the functional aspects of forming such concepts (discussed in section II). Even though in this process the idea of an ontological logic separating and combining thought and Being dissolves, this by no means implies the end of the logical production of forms. On the contrary, logic's specification of forms is expanded tremendously. In view of this constellation of loss and gain, what new possibilities for determining the function of logic emerge, and what do these prospects mean for the form of logic? These questions I will discuss in the final section (III).

I

In traditional scientific thought, the concept of form was conceived in reference to the concept of matter, which therefore figured as the basis for all form specifications. In this way unity was achieved and brought to the distinctions. This kind of thinking rested on the assumption that something, and not nothing, exists. Consequently, only two values are conceptually available to the form of logic: the value "truth," which records the correspondence with reality and must therefore be ascertainable, and the value "untruth," which serves to indicate an erroneous and corrupt controlling of "truth." A third option does not exist (*tertium non datur*). George Boole,[3] setting out to prove the completeness and systematicity of logical forms, relies on this one-dimensional ontology. He does so despite the fact that the distinction between Being and Nonbeing—as a unitary idea—no longer supports the possibility of logical forms since accessing Being itself has become a problem. There is now a gap between reference and truth, even in mathematics, so logical forms cannot be rescued by this distinction. How then do they become concepts that regulate the domain of provable propositions?

This gap is exactly the problem (and the latent issue) of modern formal logic, and it becomes the reference point for logic's functional orientation. We therefore begin with the question: How did this problem arise in the first place? I will restrict myself to one aspect of the issue, namely to Kant's discovery that concepts are grounded in the use of forms.[4]

Kant made this discovery in the context of his epistemological undertaking, which radically questioned the traditional stance that presumes the inherent referentiality of concepts. The well-known outcome was that concept-forming achievements cannot be taken for granted a priori; thus they also cannot be based on self-evident presuppositions, even ontological ones, since, according to Kant, they are achievements of reason, which is itself constrained by forms. Without getting even remotely involved in the implied epistemological questions and discussions here, we can note that the issue of the concept-forming and hence truth-regulating function of form contains the starting problem of modern logic, which initiated the development toward "formal logic." For the time being it may suffice to focus on the form analysis of formal logic to see whether and how units of form evolve. Yet at this point it is already evident that modern logic inaugurates a novel, that is, an operationally implementable, foundational program for itself, a program that, in contrast to the traditional understanding, can only ascertain the foundation of logic in and through logical operations. Let us therefore consider whether and how this program is implemented and what the consequences are for determining the form and function of logic.

II

Logic in the traditional view considered concepts, though abstractions, as referring to objects in the world; those abstractions, as mentioned earlier, are conceptualized in terms of matter or equivalent notions. This underpinning vanishes given Kant's discovery that concepts are grounded exclusively in forms. How does one then get to forms, especially to logical forms, which, as forms of truth, are references to the outside world? This question points to the form analysis of formal logic, which establishes itself as an autonomous discipline in order to ascertain the forms of provable propositions. But how can such an analysis of form succeed without necessarily relying on forms to begin with, and thus becoming enmeshed in circularities (Quine: in infinitum) or even reverting to ontological presuppositions?

In view of these traps, what moves to the foreground are methodological considerations about generating forms, and not functional orientation as a condition for contextualizing the analysis of form (which is of particular interest for our investigation). Functional orientation as a comparative approach to integration remains a "blind spot" in a self-foundation of logic that relies on deducibility and adheres to the idea that "correct" forms entail a reference to the outside world. That is why Boole employs arithmetical algorithms in his calculus to transform systematic expressions into object-oriented asymmetries with the generality of logical forms of inference (as discussed below in subsection 1). Despite the considerable increase in knowledge, however, this algorithmic treatment is not suited to help mathematics resolve its foundational problem. The subsequent attempts of logic to ground itself, still prioritizing the deducibility of forms, are hence no longer oriented toward algorithms but toward the semiotic character of form, which highlights more clearly the distinguishing quality of the form concept. Here I have Gottlob Frege's far-reaching project in mind (discussed in subsection 2). Since even this approach does not achieve the desired goal of a logically salved *Begriffsschrift* (conceptual notation), how logic's form is to be determined remains an open question. Thus, the deducibility agenda is replaced by logic's functional orientation toward problems (discussed in subsection 3).

(1) First I want to present the Boolean, so-called logical algebra, that is, a logic of propositions and predicates constructed in a mathematical-algebraic fashion. I do so for three reasons: First, because Boolean algebra inaugurated the paradigm of methodologically controlled analysis in formal logic wherein circularities are supposed to be excluded in principle; second, because it enters into an explicit cooperation with "mathematics"; and third, because Boolean logic, although constructed by arithmetical algorithms, is based on the distinction between operation and ob-

servation. Presupposing, as a second-order observation, other observations, this distinction distances formal logic from the ontological tradition and opens up the issue of connectivity. But it also remains as a distinction algorithmically held together, trusting in the workings of simplifications (techniques).[5] Nevertheless, this distinction as an observation of observations—thus as a novel form of reflection and theory—gains great importance for the subsequent fate of formal logic since it allows one to tackle the theory question of the logical enterprise in an operational fashion.

How does Boole proceed to realize this program designed for systematicity and completeness of logical forms? First, he furnishes the program with "principles": "1" as the symbol for the universe, a "calculable" reference, if you will; the idea of a class of objects as a conceptual frame; and the equation in the form "$x = x$," where x is an individual variable. These principles are forms that are certainly capable of being extended or made asymmetrical. They enable one to deduce general forms from "correct" operational procedures while disregarding the symbolic system generated by consciousness or otherwise. Hence time enters the picture, for crossing a distinction takes time. Yet this need not be reflected, since for Boole, the observer, forms are after all logical forms only insofar as they can be combined into a calculus (as a general algorithm). The reason for this is that logic is about "laws of . . . combination"[6] that stand their own ground.

Two points favor this mathematical analysis of logical forms: It seems to work successfully, and its nature is almost strictly arithmetic-procedural. Whether or not one accepts this analysis, its power lies in what may be broadly termed its "calculization," that is to say, its turning the distinction between operation and observation into an algorithm.

With this distinction, in contrast to the Kantian treatment of the ontological paradox on the level of consciousness (by means of the distinction between transcendental and empirical), logic takes on the part of observing thought by technologizing operations in order to detect how forms acquire their conceptual relevance in terms of indicating truth or untruth. In this context, Leibniz had already provided the idea of a calculus as an algorithm operating with forms (symbols), without, however, showing how this leads to the generation of provable forms. Here Boole goes a decisive step further, employing the distinction between operation and observation as a reference point for operable constructions. This does necessitate, with the sequentialization of the form analysis, the inclusion of time. But since the form analysis adheres to a strictly procedural mode without making the observer (or subject) an issue, it can be constructed and carried out without knowing the world. While establishing itself as an autonomous discipline, logic is committed to this technical method. But can logic actually do that? How is it possible for logic if the logicality of forms can only be discovered in the very act of performing logical operations? How is it that there is such an act? This rephrased question remains open. Thus, once more: How is the "generality" of forms as an indication of their provability to be expressed in a strictly procedural and hence recursive analysis?

Obviously we cannot ignore this query. This becomes clear when, in the limited context of propositional and predicate logic, Boole sets out to prove the assumed universal interpretability of the formulas he elaborated as rules for the formation of concepts. Thus he confronted a new challenge: to demonstrate how the "elective symbols," "elective functions," and "elective equations" can be related to one another in the form of a (general) calculus. To be thus related they must be not only comparable to one another but also operational in a reverse manner to fulfill the claim of universal interpretability, because the "one and sufficient axiom involved in this application[7] is that equivalent operations performed upon equivalent subjects produce equivalent results."[8] Since this problem can evidently no longer be solved in the common functional fashion of generalizations, that is, by stating what is identical, Boole understood it to be the business of a mathematician on the verge of viewing the expansion of this technification of logic as "the paradise of mathematics" (David Hilbert); although logic is hereby noticeably made dependent upon mathematics in a way that is technically no longer controllable.

(2) Here begins Gottlob Frege's undertaking to answer the form question of logic, explicitly in terms of a program of reference, by means of a "writing (i.e., notational) system" (*Schrift*).[9] In doing so he adheres in principle to the organizability of the project and, by extension, to the programmatic approach to this task. The functional ramifications of this task, however, can now come to the fore since Frege is concerned to answer the form question of logic without recourse to the algorithms of "mathematics." On the one hand, their place is now taken by the idea of

the "scriptization" of the project as the formal counterpart to the now expressly assumed content orientation of the form analysis. And, on the other hand, we find the distinction between procedure and assertion, thus the incorporation of a special level of predication for obtaining the forms of truth whereby the form analysis is referred again to language. Hence the scriptization does not prove the existence of logical forms, but serves as the plane of reference and signification for the form analysis, so that the provability of forms can be established. The "subjective contaminations of reality" are to be taken into account since it is subjectivity that "grasps what is already there."[10]

With this program Frege aims to measure the "space" between truth and subjectivity as the referential distinction in the production of logical forms. To this end he employs "writing" as a medium for combining the procedural with the referential aspect of forms.

"Writing" is accordingly summoned as an organizable medium for segmenting and combining written symbols, and thus also for sequencing operations whose reference to forms is not already established, as in the case of numbers, but first is to be discovered, irrespective of the numerical form, so that forms of truth can be obtained. Two viewpoints are hereby brought into play: the viewpoint of dissolution (segmentation) and that of retaining (recombination). In this respect writing works as a kind of blanket form open to the construction of units. Thus writing is a difference; its distinguishing operations can be used to represent logical forms.[11]

This description of writing, capturing the aspect of unity yet not constituting the unity itself, becomes fruitful for the form analysis of logic since it enables an inquiry into the logicality of logical forms in terms of what constitutes the unity of an operation. This inquiry begins with the distinction between procedure and assertion in order to include the use of form itself in the form analysis of logic—that is, in order to grasp and describe the use of form, or to "discern" logic's form by way of the consequences of the form-generating process.[12] Thus for Frege the question of the logicality of logical forms is not a matter of intuition or special certainty. How then do we gain insight into the logicality of forms?

But first: How is this analysis of form set in motion? By means of a connecting symbol to introduce an element of limitation into the relative arbitrariness of "writing." This limitation in turn is rendered by a symbol for "conditionality" bringing to bear the aspect of connectivity. In Frege's system, connectivity is materialized by way of the symbol for so-called material or factual implication, which, however, does not yet indicate a relation but rather marks a junction whose truth content is still indeterminate.[13] That is why Frege needs the distinction between procedure and assertion: to make room for the possibility of predicating the truth content as the consequence of such scriptization. But how? By including an observer, whose possibilities, however, are at first still bound to the propositional structure of affirming or negating truth values in terms of the traditional *tertium non datur*.

Prima facie this analysis of form seems to amount to nothing but reproducing tautologies. But why "nothing but"? Because, by introducing further symbols that themselves encounter previously introduced symbols, and thus distinctions, Frege notices with obvious surprise that within the evolved writing system symbols "at once appear *in propria persona*."[14] What has happened?

Frege's analysis of form does not rely on the correctness of the mathematical algorithm. Moreover, it does not rely exclusively on the combinability of the introduced and designated symbols warranted by the propositional structure, but also depends on a specific content level to have the truth values operationally available independent of the classic linguistic instrument of predication. This content level, though, becomes, or is, only relevant to the search operations of the form analysis insofar as content can be assessed.[15] If this content (however it may be designated) can be judged, logically relevant concepts are formed on the condition of connectivity as long as the combinations, however made and judged, can be related to "content *equivalents*." What does this figure of equivalence now mean for the form of logic? If "equivalence" becomes the trademark of logical forms, how then can this form be determined more precisely without the inclusion of the propositional structure?

Frege replaces the distinction between subject and predicate with the one between function and argument.[16] This is an abstraction that is regarded and dealt with as operable and generalizable and that therefore has a conceptualizing effect. That way, it is true, the analysis of form is freed from the subject-centered form use of the propositional structure. But how then can the form use be salvaged as a logical one? By applying it. For that, it is necessary to relate the "conceptual notation" to specific productions of form so we can see how and with what results Frege's symbolic language, for example, establishes arithmetic, since "arithmetic is merely a more highly developed logic."[17]

Arithmetic is understood as a set of numbers; it always contains statements referring to a concept, implying that they are also determined by a concept. The enterprise of "applying" the "conceptual notation" does not consist in "counting" but in tracing back the arithmetical operations to "basic laws" of logic (axioms) that are necessary to define these operations formally, that is, by demonstrating the procedures ("general" algorithms). Since this undertaking has to face the notion of "infinite" quantity—still understood by Frege and Cantor as a number of a countable infinite set, the conceptual notation is now challenged from the form side, which hitherto has only been noted (as a vertical line without a proper symbol) under the heading "content." From this point Frege further develops the "first" notational system by introducing both a symbol to indicate objects in the form of a definite article and a symbol to designate the value performance of a function. Only through this conversion of the "content" can one begin to deal with and define the "power" of numerical sequences in a formal-logical fashion that mathematicians of the nineteenth century had already attempted to master. And it is at this point that the "Frege catastrophe" occurs, since this formal-logical representation of arithmetic along the conceptual sequence (namely: number, defined as the extension of the concept, defined in turn by the value performance of functions) inevitably enables the construction of logical contradictions in terms of the presupposed binary classical logic. This possibility, however, cannot be "localized" within this type of production of logical form.[18]

What has happened? In reframing "content equivalence" by means of the distinctions that now also pertain to object reference, namely unmark/mark and thought / truth value, the attempt to achieve a universally provable logic was stamped with the "impress of paradox."[19] However, the methodological distinction between object and concept was retained as the frame for the analysis of form, despite the fact that the conceptual notation was proven incapable of connecting this distinction to others without contradiction. Or again, from the viewpoint of form as unity and distinction: the classical logic of the *tertium non datur* appears in this representation in a paradoxical (circular) closedness, thus with an inherent contradiction. Logic must therefore be conducted as the conditioning of the contradiction; that is, as a system of references when applied to the notion of horizon as a metaphor for the world, or as the "infinite" repeatability of the same when applied to mathematics.

(3) Here Frege hoped for a solution, but he did not find one because, constrained by traditional logic, he continued to search for finite solutions. As a result, for Frege logical research ends with Bertrand Russell's discovery of antinomy in set theory (1901). However, logical research does not end here but responds in a specific way, that is, with reference to contradictions. And it does so in the broad sense of searching for solutions to the "unsolvable" problem of the unity of form,[20] which implies that this approach on the one hand follows Frege, yet on the other hand does not. The Fregian line is continued insofar as one still assumes the ability to apply the same operation repeatedly to the outcome of the previous operation as the condition for comparing and extending logical forms. But since this form of recursive operationality as the trademark of logical (and mathematical) operationality by no means excludes the unfolding of contradictions, the guiding ontological difference between thought and Being linked to classic logic is transformed into the difference between observation and operation. Consequently the "broad" functional orientation becomes the determining factor emphasizing the extension of comparability and the like, and proving itself in generating questions.[21]

But insofar as the analysis of form continues to adhere to the position of deducibility, it is preoccupied with satisfying the demands of this viewpoint, however reduced, with respect to the validity of the formulae produced. Consequently, in the aftermath of the Fregian blockade, many different procedures enter the scene: one responds to the universal attitude toward axioms with the "axiomatic method," to the questionable quality of the *tertium non datur* with its decoupling from ontology, and to linearization and sequencing with the insertion of hierarchical decision structures. In short, procedures and distinctions are developed to contain the now-virulent functional viewpoint within the traditional understanding of logic's functional orientation.[22]

After Frege, horizontal (one-dimensional) representations of logic's form were replaced by models of steps, layers, or linguistic levels. But what follows for logic's own analysis and use of form? At first this development leads at least to the structural option of observing how the form (of logic) gets to the concept. That, mind you, is an option that forgoes asking the reference question. Henceforth, this question is handed over to "the empirical domain" as a problem of validity. Still, one remains "close" to the problem of logic's form: How are "general" concepts nevertheless generated on the basis of the repeatability of operations? Indeed the analysis of form "can do it." Yet this way it runs into altogether new, that is, "independent," problems. These problems, like those of undecidability, incompleteness, and contradiction, can neither be reduced to one another nor otherwise solved within the context of recursively generated encodings without committing a sin, as it were, by using a predetermined semantics, be it even the intensional semantics of concepts.[23] This does not, however, exclude insights into far-reaching equivalents.

Judged by its results, the analysis of form, initiated by Boole as a methodological enterprise, has thus reached the status of organized circularity. But the last word on the function of logic has not yet been spoken. Explicitly including this issue of determining logic's function in the analysis of form, I think, advances the form analysis further.

III

With Kant's discovery that concepts are grounded in the use of forms, the problem concerning the unity of form was posed but certainly not fully formalized. Spencer-Brown and his calculus of form set out to tackle this task with respect to the distinctions a concept of form presupposes.[24] This calculus of form, however, does not yet take care of the business of either logic or mathematics. Yet the calculus's concept of form as two-sided does indicate why the logical analysis of form, attempting to dissolve the problem of form (as the unity of the concept of form and that of matter or content) into distinctions, must ultimately fail: All achievements resulting in unities are based on fundamental paradoxes.[25] This is why what ultimately matters for a form analysis of logic that attempts to generate logical forms out of operations (distinctions) is an arrangement containing a symbol for negation that, independently of the ontological prerequisites of the *tertium non datur*, allows the formation of concepts to regulate provable propositions.

As described earlier, Frege's arrangement proved contradictory in its "application" of the "conceptual notation" to mathematics,[26] namely, in its deduction from logic's "basic laws," without the possibility of localizing, that is, differentiating, the contradiction *within* the arrangement. Consequently, the universal control-function of logic, to the extent that it rests on the classical principle of either/or (*tertium non datur*), was blocked. This outcome prompted us not only to look for a new functional orientation for logic, but also to realize that such an orientation is already at work in the form analyses of modern logic: namely in the shift from the functional aspects of generalizability and sytematizability of forms to the aspects of comparability and reliability. Logical research, to be sure, is not hereby decoupled from the binary schematism of truth and untruth, yet contradiction has become a self-generating fact the possibility of which cannot be eliminated. That is why a formalism's "freedom from contradiction" is conditional as well. And it may well be important to refrain from proving this point, as, for example, in the known cases where the proof of formalism's completeness or questions of decidability are concerned.[27] Thus it becomes clear that logical research itself is concerned not only with eradicating contradictions. For that reason the notion of contradiction, especially where it supersedes the ontological view of predication, acquires an autonomous and central role.

From a sociological vantage, this is now the starting point for the "broad" functional orientation that encompasses the level of (internal) operations and serves as a guiding perspective, since the dependence of the unity of operations (including logical objectives) upon triggering conditions cancels out the idea of a general deducibility of truth forms. Its place is now taken by the concept of redundancy,[28] along with the functional orientation of comparability. The notion of redundancy denotes abundance and dispensability, features that enable enlarging and subsequently limiting the scope of potential connecting knowledge.[29] Thus redundancy designates security techniques in information processing (informational redundancy) via communication, for instance, repetitions that, as isomorphisms, replace ontological assumptions of constancy. Once redundancy supersedes deducibility for the purpose of forming (regulating) provable communications (propositions), the question ensues as to how the form of logic is then determined.

If, with this functional shift, truth interests are replaced by security interests without decoupling the latter from the binary schematism of truth, and if consequently the notion of contradiction as a self-generating fact acquires key significance, then how can the form of contradiction be described without relying on the "trinity" of classical logical principles?[30] Further, how is the concept of contradiction as a form determined? Looking at Frege's analyses of form, it becomes clear that the form of contradiction is correlated with the entirety of the logical formalism we have presented. But since the possibility for contradictions cannot be localized on the operational level, this question is suspended. Here the functional orientation aids us again insofar as it is focused on securing the connectivity of the operations. As Luhmann proposed, the notion of conditioning signifies those conditions or prerequisites:[31] This is so because the ability to condition the relations among elements, that is, to limit them with respect to specific functions, secures their redundancy as a conditioning function.

In this sense we have already tacitly employed the notion of conditioning when, while describing the analysis of form, we pursued the questions regarding the conditions that set operations in motion. But not only then. For in general this concept of conditioning enables observing specific formations of concepts that have to cope with the problem of the two-sided form. And this then constitutes the special challenge in forming logical concepts that are meant to deal methodologically with "truth."

Thus, regarding the functional determination of logic outlined above, the concept of redundancy denotes a *conditioning function* such that the question arises: How are logical forms conditioned? Or in a similar vein: How are logical forms determined, given that they are not suited to control processes in a manner free of contradiction. Answer: They are meant to produce *and* recognize contradictions.[32]

Logical forms (and likewise logical calculi) thus are techniques for constructing and identifying contradictions. How they may be eliminated, for instance, without the loss of information or other "substance" (content), depends on the type of science involved. But the sciences must confront this question and indicate their respective solutions if they claim to control knowledge.[33]

But how, in contrast to the traditional usage of the binary schematism of truth and untruth, is this functional determination brought to bear on the logical formalism itself? Thus how is this form as a redundancy-containing form generated and secured operationally?

If, according to the functional conceptualization, logic is concerned with the generation and resolution of contradictions and hence with monitoring redundancy in the sciences, then it has to represent itself as an order containing redundancy, not as a deductive order. Accordingly, the representation of formalism rich in structures is confronted not only with the demand for freedom from contradictions but especially with the demand for decidability in scientific statements. This must and, indeed, can be secured, however fictitiously, with respect to the assumed coherences.[34]

Therefore, logical forms are determined as redundancy-containing forms whose unity in principle rests, however precariously, on equivalence as the standard point of reference. In programmatic terms, that is, as structures of organized connectivity, they fall under the conditions of redundancy, which necessitates that the idea of how to apply logical forms be shifted from deduction to argumentation guided by observation. And for this undertaking, in the end, the distinction between operation and observation as a form of theory guided by functional thinking is recommended.

ELENA ESPOSITO | **TWO-SIDED FORMS IN LANGUAGE**

I

According to Niklas Luhmann, form signifies a distinction "provided the distinction is distinguished from what is distinguished by the distinction."[1] The circularity of the whole approach is already apparent in this initial definition: The point of departure is a distinction that can only function as such on the condition of another distinction, namely the distinction that makes determining the first one possible. As Spencer-Brown puts it,[2] the starting point is a *mark* that distinguishes both of its sides (the *marked state* and the *unmarked state*), yet it does so against the backdrop of an *unmarked space* from which it distinguishes itself as form.

This seeming inconsistency—together with the fact that the starting point of the construction (as if it could be free of presuppositions) presupposes something itself[3]—has evoked numerous critiques of Spencer-Brown's approach.[4] Without detailing the controversy here, we can note in all of the criticisms the shared belief that Spencer-Brown "missed" a point and therefore his construction must be supplemented. Apparently it is assumed that he was not fully aware of the circularity implied in his construction and that he "forgot" to include himself and his observational perspective. According to Francisco Varela, for example, to include oneself and one's perspective requires that the *calculus of indication* be modified into a *calculus of self-reference* in which the operator implies the reference to itself.[5] The sign for distinction ($\neg$) then becomes the sign for uroboros ($\square$), signaling, by referring to itself, an awareness of the reference to something other: Thus, aside from the distinction (of form), the distinction of the distinction (the form of the form) enters the picture. It is itself a two-sided form always implying the reference to what is not distinguished, that is to say, in Luhmannian terms, the reference to the form's outside.

Beginning with the form's form, however, does not yet constitute a solution. The objection that the reference to further distinctions repeats itself infinitely can be countered by pointing to the properties of self-reference, but, given a multitude of observers, this counterargument is not quite convincing. To somewhat simplify the matter: If one proceeds from a particular observational perspective, nothing justifies the decision to privilege the perspective of the self-observing system over any alternative—let alone the fact that this system can observe how it is observed by others and that it can also multiply the distinctions of distinctions from its own perspective *ad libitum*. If instead one goes back to the specificity of operations (ultimately to autopoiesis) characteristic of each system individually, then the infinite regress is stopped, but it is no longer clear why the "form's form"—which pertains not to the operations but rather to the way operations are observed—needs to be taken into account at all.

On the other hand, we have to note that the awareness that form itself can be (or must be) distinguished from something else is explicitly present in Spencer-Brown as well.[6] He suspects that any form (any *cross*) is enclosed by an "unwritten cross" indicating in each case the ever-present potential for discerning form as form. This latent *unwritten cross*, one could say, is nothing but the shadow of the *unmarked space* within the construction proceeding from this space. There is thus an implicit reference to the paradox constitutive of Spencer-Brown's construction.

The situation is thus more intricate than was initially apparent. In order to re-include the position of the observer in a circular construction, it is not sufficient to insert another level of distinction. How else can it be explained that Spencer-Brown himself, though aware of the problem, did not make use of a self-referential form?

These sorts of questions are the point of departure for the following considerations. It is important, I believe, to accentuate the fact that in Spencer-Brown's calculus the fundamental operation itself (namely the operation of distinction/indication) is of a complex structure. Once this point is clear, it becomes possible to pose the question regarding the form of the form more clearly and consequently to better recognize the connections between the problem of self-reference and the one of observation. Against the backdrop of such a clarification we can tackle the matter of analyzing the form of language.

The decisive point, not always sufficiently stressed, is that Spencer-Brown's calculus is an "operational" calculus: It refers to the concrete operations of a particular system. Asymmetrization, which prevents entrapment in the distinction of the distinction of the distinction and so forth, is primarily sought not in self-reference but in connectivity, that is, in the fact that the operations of the system continue. First, I would like to distinguish indication/distinction as an operation from what is indicated/distinguished: Every operation of indication indicates something, but indicating as an operation is not exhausted by what the indicated object is in each case. Each operation distinguishes something to which it refers, yet at the same time it generates the distinction between the operating system and that to which the system refers. These two distinctions are not congruent: The distinction between the object indicated in each case and that from which it is distinguished does not match the one between the operation of the system and what is external to the operation. We shall address the latter as the *self-reference / external reference* distinction (s/e), in contrast to the distinction *indication / distinction* (i/d) that guides the operations. The circularity of the construction is revealed in the fact that distinction is a case of self-reference (distinctions only occur in a system), whereas indication belongs to external reference (the indicated objects do not coincide with the ongoing operation).[7] Form, strictly speaking, implies the relationship between the distinction i/d and another that implicitly always exists: the distinction between self-reference and external reference. When we examine the relationship between those two distinctions (or their being distinguished), we are concerned not simply with distinguishing two forms but with the "orthogonal relation"[8] that rests on the possibility of treating a form as form (that is, a form distinguished from its two sides). As I will show later, this is equivalent to the potential of employing form for observations.[9]

Hence, to whatever distinction form will be related, a distinction between the distinctions i/d and s/e must already have been made. Although these two distinctions pertain to two different levels of articulating form, they do presuppose each other: one cannot be had without the other, yet they remain distinct. The distinction between the distinction itself and what is thereby distinguished rests on this phenomenon. To give a linguistic example: One can only talk about an apple as being distinct from other objects (i/d) if apple the object is not taken for apple the name, that is, if the object is not confused with the name of the object (s/e). On the other hand, the distinction between object and name (s/e) evolves simultaneously with the ability to designate the corresponding object (i/d). The two distinctions are simultaneous and equally fundamental even though they are not equivalent: Distinguishing one object from another or one name from another is different from distinguishing objects from names.

Following Spencer-Brown, one could say that the indication/distinction distinction guides the ongoing observation that always indicates an object as something being distinguished from something else (i/d). Observation is always of something, and the very possibility of directing an observation toward an object presupposes the object's simultaneous distinction from something else. The distinction then can exhibit different features requiring more or less sophisticated assumptions: It can be a simple distinction between the object and an indeterminate "other" (this / not this; apple / non-apple); it can be the specification of an opposite (woman in contrast to man, and not just to non-woman or animal); or finally it can be a technically elaborated binary schematism excluding third values (true/untrue, justice/injustice, etc.). In each of these cases it is necessary to indicate one side against the backdrop of the other that is implicitly negated. Both sides, however, are located *within the system* since distinctions exclusively exist in a system. In the environment one can find apples and women but no distinction between apples and something else or between women and men. The observing system always operates with its own distinctions even though they are projected onto the environment.

Thus the relationship to the environment does not concern the i/d distinction, which, so one might say, presupposes on the part of the observing system the ability to keep apart objects from observing those objects. In other words, this distinction presupposes the separation between signifier and signified such that observing an object is not conflated with the object observed: Observing an apple is something different from eating an apple.[10] Or: Observing an observer is something different from communicating with that person, and that is why self-observation exhibits special features. The relation between inside and outside, by comparison, pertains to another distinction—namely the one between self-reference and external reference—which accompanies any use of indications or distinctions without exception. The s/e distinction pertains to the way the reference to the external is made within an observing system or, more generally, to the issue of reference. In the case of language, this distinction concerns the relation of words to their extralinguistic counterparts.

A constructivist approach obviously cannot proceed from an independent external referent. Here the question of reference pertains to the way the observer, in observing observations, distinguishes between self-reference and external reference. The issue explicitly comes up only in the case where an observation (which must use an i/d distinction) utilizes the s/e distinction to indicate one of its sides in contrast to the other side. It is thus a special case of observation with particular presuppositions. Only in this case do the two distinctions coincide, a situation that generates quite a host of problems—in the domain of linguistic theory, for instance, the problems surrounding the distinction between using and mentioning words.

Apart from these cases, however, the distinction between external and internal reference is always reproduced by observations and always intersects the i/d distinction. Both distinctions proceed relatively independently of each other (they are positioned "orthogonally" with respect to each other): The observation can be directed toward either self-reference or external reference by denoting objects in distinction to other objects or words in contrast to other words. Characteristics of one side do not affect characteristics of the other. (The word "fire" does not burn.) They also work differently. Whereas i/d-type distinctions use *negations*—thus separating two equal and opposing sides—self-reference does not negate external reference (and vice versa): The object is not the negation of its name; it is only the referent for the name. Moreover, the relationship between self-reference and external reference is *not symmetrical*: The distinction is always drawn within the system.

Be that as it may, within the general framework for constructing this theory it is not merely accidental that the two distinctions and their simultaneity play central roles. They do so because they are connected to two fundamental principles of the autopoiesis of observing systems (and from a constructivist viewpoint, only systems can observe): one is the *autonomy*—the differentiation—of the systems within an environment (s/e distinction), and the other is the *connectivity* of the system's operations (i/d distinction). A system can only reproduce itself as an autonomous domain of operations as long as it succeeds in securing for every operation the necessary connectivity to generate subsequent operations. That constitutes a coding problem: How does one get from one word to the next? Operations can be connected with other operations of the same system only if the required distinction between inside and outside is maintained (i.e., if objects are not mistaken for the objects' names). That in turn is a problem of reference. Once again we are dealing with two different (though interdependent) problems.

From what has been said thus far, we can define a two-sided form as the distinction between two orthogonally related distinctions, each of which pertains to several aspects of the autopoiesis of observation:

(1) the s/e distinction

- autonomy
- problem of reference
- asymmetry between observer and object observed

(2) the i/d distinction

- connectivity
- problem of coding
- symmetrical use of negations

According to Gotthard Günther, the relationship between the two distinctions can be regarded as an instance of a "proemial relationship,"[11] that is, as a distinction between a "symmetrical exchange relation" and an "ordered relation." It is "proemial" because it constitutes the necessary precondition for making distinctions and thus for generating forms. Such a relationship represents the basis of the very possibility of connecting a relator to a relatum (an observer to an object) under conditions of self-referentiality, always implying four "relata" (thus two distinctions) at the same time. We are, in other words, confronted with a complex relationship from which emerges the possibility of observing objects without assuming their independence from observation.

II

How then do forms, as specified in the previous section, work in practice? How does one detect two-sided forms in particular observations?

What does the analysis of form achieve where inquiries of specific observations are concerned? Does it only assert that form is involved (as in any observation), or can the analysis also yield precise information on the type of observation and the way it works? Is the notion of form an operational concept? In what follows, the case of language will be explored with regard to both the interconnections and the simultaneity of the i/d and s/e distinctions.[12] We assume that we must have language at our disposal to be able to employ two-sided forms and that the terminology developed in linguistic theory, once interpreted appropriately, can furnish us with useful insights into the properties and condition of such forms. One can then seek to determine in other cases whether and how nonlinguistic forms display the same properties and what conclusion can be drawn from such findings.

So that we can elucidate the reasons for language's central position, our investigation will proceed from ideas found in Niklas Luhmann's semiotic theory.[13] Luhmann develops a concept of sign that views the sign as a form of a particular distinction: the distinction between signifier and signified. A connection with traditional investigations of the signifier/signified (s/S) distinction[14] is thus established, namely with linguistics' prominent notion of the "two-sided unit" ("unité à deux faces").[15] Luhmann's theory, however, fundamentally differs from this tradition in that for him signs are explicitly related to the operations of systems, and hence the inquiry of "semeiosis" is subsumed under a theory of operationally closed systems (psychic as well as social). As one obvious consequence of this subsumption, the idea of the external reference of signs is abandoned.

The central questions in this approach address the relations between signs and the operations of the corresponding systems, and especially the conditions for a system to be closed. How can an operationally closed system, which lacks any contact with the outside, relate to the s/S distinction? How are we to interpret the notion of the signified, that is, the reference of signs? And, above all, what is the purpose of reference?

In this area Luhmann also offers suggestions. The working condition for signs consists in a particular combination of isolation and redundancy. By *isolation* Luhmann means that the relation between signifier and signified needs to be decoupled from any reference that is considered independent of the pure distinction of the two sides: The word "apple" denotes the object apple simply because it is the other side of the word/object distinction and not for any other reason. The word bears no similarity to the object; it is not connected to the object by any rules; and it does not inform us about the appearance or the behavior of the corresponding object. Thus references between signifiers and signifieds are purely intralinguistic and do not depend on any other connections of meaning. Isolation as a feature of language clearly resembles Saussure's notion of the arbitrariness of signs, but in contrast it emphasizes the connection to operations of meaning-based systems. We are dealing with operational closedness: A domain becomes differentiated, to which special conditions, decoupled from the conditions of the outside world, apply.

The isolation of operations, however, is not the sole condition necessary for the reproduction of an autopoietic system. A system exists as long as it is able to replicate its own elements. It succeeds when each operation successfully joins another operation of the same system and proper connectivity is secured. In the case of sign operations, each sign must contain information for subsequent signs yet remain arbitrary. That means that this information cannot depend on properties of the signified objects (as if an apple "intrinsically" would refer to a pear rather than to a book) because those properties do not belong to the s/S distinction. Every sign must exhibit a certain amount of *redundancy*; it must contain a surplus of information on the subsequent signs and thereby limit the range of possible connections. This information, however, must pertain to the sign qua sign (as a "two-sided unit") and not to just one side. References are thus not references between pure signifiers: The sound "apple" does not impart any information about subsequent sounds.[16] What counts is only the information contained in the "signification" of the apple, which does not exist anywhere outside of signs.

If we accept this approach, and we require both isolation and redundancy of signs, then a first difficulty arises: It becomes unlikely that the distinction between the two sides of form will be maintained in the course of the operations. Redundancy secures the iterability of signs such that the s/S relationship need not be reinvented in each instance. Isolating the distinction, then, is arbitrary not in the sense that it is free of all connections but only in the sense that it exclusively depends on connections within the network of signs. In using the word "apple" we thus refer to a specific meaning that is determined by prior uses and by the redundancy of other words employed in this context. Every time the same signifier is used, the identity[17] of the corresponding signified is confirmed. But does this regularity not jeopardize the "two-sidedness" of the sign? If a signifier regularly corresponds to the same signified, it stands to reason that in the signifying process a correlation between the two would inevitably evolve over time and that the properties of the signifier would gradually adapt to the nuances of the corresponding signified.[18] Such a development can be observed, for example, in language's prosodic features[19] (intonation, tone of voice, etc.): Without knowing a particular language, it is possible to recognize by tone of voice whether a conversation is friendly, puzzling, professional, and so forth, precisely because the relationship between phonetic signifiers and the emotional signified exhibits a certain isomorphism.[20]

If this were to happen, the isolation of the signs would be jeopardized because references of meaning would be connected with "natural" characteristics that were partly independent of the signification as such. The differentiation of the distinction between self-reference and external reference would be undermined, and consequently so would the autonomy of the operations.

What precludes such a development? Can we discover in the way operations are connected (i.e., in redundancy) properties that prevent a flattening out of the two-sided semiotic form into a *unité à face unique* (unity with a single side)? According to the thesis we have set forth, these properties coincide with the special features of linguistic signs that set them apart from other types of signs. Even though the definition of the two-sided form does not focus on language, reproducing and maintaining forms with two "planes" in a closed context requires certain conditions that are most effectively realized in language.[21]

III

But how can language reproduce both isolation and redundancy at once? Which are the properties other "sign systems" lack on which must rest the ability to utilize two-sided forms without letting them collapse into "units with a single side?"

Since Saussure, the *arbitrary nature of the linguistic sign* has come to mean the absence of any "natural connection"[22] between the signifier and the signified: "The idea of 'soeur' is not linked by any inner relationship to the succession of sounds *s-ö-r* which serves as its signifier in French."²³ The conception of the arbitrariness of signs provoked an intense discussion that, in my view, has time and again produced misunderstandings. In an often-cited article, Emile Benveniste raises the objection that for the user of a sign the relationship between signifier and signified appears to be absolutely necessary: "The concept (the 'signified') *boeuf* is perforce identical in my consciousness with the sound sequence (the 'signifier') *böf*."²⁴ According to this view, calling the signifying relationship arbitrary is only appropriate in those cases where no connection whatsoever exists or, according to Roland Barthes, in those cases where the sign is "artificially" chosen by a group of people who create it on the basis of a "unilateral decision."²⁵ However, if arbitrariness in this sense is equated with free choice (*arbitrium*), then what is essential in Saussure's approach is missed: Saussure explicitly distances himself from the idea that the signifier depends on the free choice of the speaking subject.²⁶ The sign is obviously "imposed" on each speaker. For Saussure it is rather important to highlight the autonomy of the signifying relationship, which is not contingent on factors external to the relations among signs. Using an expression that would later enter into the technical jargon of linguistic theory, Saussure maintains that signs are first of all "unmotivated."²⁷ It is arbitrariness that allows us, for instance, to speak of both the mutability and the immutability of the sign:²⁸ Precisely the arbitrary nature of the sign—together with the system's complexity—"protects language [*langue*] from any attempt to modify it. . . . The reason is simply that any subject in order to be discussed must have a reasonable basis."²⁹ Since, however, the sign is arbitrary, nothing can stop language from continually changing and from shifting the relationship over time between the signifier and the signified.³⁰ Both the impossibility of an external intervention via signification and the constant alteration based on language itself are nothing but a confirmation of language's fundamental autonomy.

Thus far these insights from linguistics do not exceed what second-order cybernetics or radical constructivism have to offer:³¹ Once the idea of a univocal world of reference is abandoned, it no longer makes sense to speak of arbitrariness *per se* but only of arbitrariness with respect to an observer. If "arbitrary" is taken to mean that "given the form, it is impossible to predict the meaning and, given the meaning, it is impossible to predict the form,"[32] then it follows for the theory of observation that the relationship between form and meaning (or signifier and signified) cannot simply be obtained from the objects. Rather it must be related to a special observer, that is, to a system that constitutes the distinction between the two sides of the form concerned. The s/S relationship is a relationship constituted by a particular system and is internally produced and transformed on conditions of closedness. This is but one aspect of the system's autonomy. From this viewpoint, autonomy is never the problem of the first-order observer (the one who is using the sign) but only a problem of the second-order observer, who examines the conditions for observation and relates them to the operations of the system.[33] It does not make sense to speak of arbitrariness universally: Either the question immediately arises, "arbitrary to whom?" or one assumes a position similar to Benveniste's in which the different observational perspectives are not distinguished.

Even if arbitrariness is not related to an observer, it is not at all equivalent to randomness of connections. The linguistic sign, as Claude Lévi-Strauss asserts,[34] is arbitrary *a priori* but not *a posteriori*: The arbitrariness at the beginning is absorbed in the course of the system's operations by the recursivity of those operations.[35]

Why then is it of interest at all to make arbitrariness an issue? What does one gain from starting with the arbitrary nature of the linguistic sign? The relevance can only be judged by the implications for the construction of our theory. If "arbitrary" is taken to mean "nonmotivated," then "arbitrary" is synonymous with "non-analog."[36] Ratios are called analog when a proportional relation between two series of entities are assumed (s/S),[37] that is, a relationship that necessarily implies a certain isomorphism (or "similarity"). But the negation of "analog" is "digital," meaning "based on binary oppositions."[38] As Barthes asserts, binarism is the purest form of digitalism.[39] In other words, if the sign does not imply isomorphism between signifier and signified, then it must rely exclusively on distinctions, and distinctions, as discussed earlier, exist only within systems.

If this is true, and if the principle of the arbitrariness of signs needs to be taken seriously, then it ought to be possible to reconstruct the entire linguistic apparatus solely on the basis of binary oppositions. The work of the Prague School can indeed be regarded as an attempt to implement such a program, on the basis of the prominent principle of binarism. Starting from the acoustic plane, phonology reconstructs the relationships among sounds in terms of articulations of different types of opposition[40] that can also be found on the verbal plane. Each linguistic element acquires its validity by its insertion into a network of oppositions to other elements of the same language—and not because of any assumedly independent content. This is evidenced by instances where "content" becomes or remains absent while still performing a discriminating function: for example in the well-known case of the "zero sign,"[41] where the absence of an explicit signified functions as the signified. It corresponds to the "absolutely differential value of the language."[42]

In this case as well, Saussure is the point of reference,[43] especially his concept of *value*.[44] Linguistic elements, he maintains, "are purely differential and defined not by their positive content but negatively by their relations with the other terms of the system":[45] "In language there are only differences."[46] The identity of words hence is not dependent on inherent properties but exclusively on their "value" within the system; thus "language is a form and not a substance."[47]

A "form" in this sense is an autonomous system independent of its environment with respect to the identity of its constituting elements. Being arbitrary and being differential are two correlative properties[48] that, however, need not be congruent. If the relation between signifier and signified were not completely arbitrary, the concept of value could not be held up in its purity since considerations would enter the picture that are external to the pure distinction of linguistic units.[49] Conversely, however, if the system did not rest exclusively on interconnected distinctions—and this is the crucial point here—it could not possibly maintain its autonomy vis-à-vis references of meaning being decoupled from indications (isolation). While arbitrariness directly pertains to the relation between signifier and signified (the s/e distinction), language's differential nature relates to language-internal distinctions dealing with relations among words: These are distinctions of the indication/distinction type that cannot be attributed to relations between inside and outside. Here, in the specific case of language, we rediscover the orthogonal relation between the s/e and i/d distinctions from which we started. If the elements of language were not exclusively defined by distinctions between indication and distinction, the s/e distinction (s/S) as a two-sided form could not exist.

The "two-sidedness" of language as form is thus secured by the fact that it is first and foremost a "domain of articulations, and the meaning is above all a cutting-out of shapes."[50] The fundamental problem of linguistic theory is therefore how to reconstruct the articulation of distinctions upon which the constitution of meaning is based.[51] And it is precisely the articulation that marks the uniqueness of language in contrast to any other "sign system." As will be demonstrated later, the analysis of articulation brings the relation between the two distinctions s/e and i/d center stage. One consequently is able to ascertain what constitutes their relevance for linguistic operations.

IV

The key notion is, then, *double articulation*. This notion, introduced by André Martinet,[52] signifies the particular fact that language exhibits two separate planes on which units are articulated:

- a *first articulation into significant units* (the monemes or morphemes) correlated with one meaning ("unités à deux faces"),[53] and
- a *second articulation* into *distinctive units* (the phonemes, Hjelmslev's "figures"): simple segments without meaning ("unités à face unique").

On the first plane, units with a specific semantic value are distinguished that often (though not always) correspond to single words. A moneme is "the smallest segment of speech to which one can attribute meaning."[54] A statement such as "I have a headache" can be broken down into four segments: *I*, *have*, *a*, and *headache*. The entire statement is a sign; it is thus a signifier to which a specific meaning corresponds. Yet at the same time it also consists of four relatively autonomous signs, each of which can occur in very different contexts but nonetheless remains recognizable and meaningful. Each sign itself is therefore an instance of "unité à deux faces," that is, a two-sided form.

The possibility of breaking down speech into relatively independent segments constitutes, as Martinet emphasizes, a fundamental "economic" principle: Theoretically at least, it is possible to generate infinitely different linguistic signs—corresponding to many different contexts—without needing a specific signifier for each one. With a finite number of words, an infinite number of statements can be produced. This is one aspect of language's *productivity*, one of the properties that constitutes its specificity: namely the property allowing the construction and interpretation of new signals, that is, of signals that have never occurred before and that need not be part of a user-accessible list of preexisting signals—as extensive as it might be.[55]

Economy alone, however, does not suffice to explain language's productivity. Some form of iterability of signs is certainly needed, yet it is important as well that the components resulting from the "segmentation" of signs are general enough to be able to adapt to different contexts time and again. Every word has to be specific enough to be recognizable (as something identical), yet nonspecific enough to be somewhat different in each instance since the context always varies. Regarding the latter condition, we encounter once again the requirement of the arbitrariness of signs, which can only be met by signifiers that do not have any motivated connection to their corresponding meanings. Motivation would block the requisite flexibility of signs.

For language to bear the property of productivity, it must, therefore, be both economical and arbitrary; it must simultaneously display redundancy and isolation. But these two requirements could not be fulfilled without the aforementioned double articulation, which becomes, especially in this regard, absolutely essential. Double articulation, however, is a property only to be found in language.

On the acoustic plane, each word (that is, each unit resulting from the first articulation) is segmented into a series of distinct units: the phonemes. For example, the word "head" consists of the three phonemes "h," "e," and "d" (in this order). Unlike the monemes, the units resulting from this segmentation have no meaning: They are "one-dimensional" units, not two-sided forms, and hence not signs.

Just as an expression consists of units each of which has an identity outside of a particular context, so the monemes are composed of phonemes whose identity is the outcome of a network of distinctions among phonemes. They are always purely negative units that are, though not defined by their extralinguistic "acoustic substances," nonetheless units not solely dependent on any particular context: The phoneme "t" can be recognized as "the same phoneme" in the words "table" and "seat." The benefit resulting from the first articulation thus recurs on the plane of the second articulation, this time, though, with an additional enhancement typical of reflexive mechanisms:[56] it is a case of segmenting a previous segmentation. The first articulation enables the creation of infinite sentences starting from a finite number of words. However, the first articulation nonetheless requires a few thousand different monemes. If each of those monemes was an indivisible unit, handling each unit with all its relationships would be extremely burdensome in every instance. If each unit were to correspond to a unique identity on the signifier level, the limits of what could be expressed would be quickly reached:[57] One cannot imagine actually having at one's disposal those thousands of words that exist in every language today. The monemes, to be sure, can be broken down as well, and this time the number of autonomous units is limited to a few dozen. With fewer than 30 different phonemes, a variety of words—and a potentially unlimited variety of sentences—can be generated.

Yet the parallelism between the first and second articulations conceals a fundamental difference: while units resulting from the first articulation have both a signifying and a distinguishing function, phonemes serve exclusively to distinguish. Their identity only depends on their difference from other phonemes and does not correspond to any meaning.[58] Once again, phonemes are "unités à face unique."

This property, however, does not pose a problem. It is to be considered jointly with the second condition for language's productivity, that is, the nonmotivation (isolation) of signs that constitute language. "The phonemes, products of the second linguistic articulation, thus turn out to be the guarantee of the arbitrariness of the sign."[59] "If the form of each moneme were an unanalyzable mumble, there would be a complete joining [*solidarité*] of meaning and acoustic form."[60] As indicated above, nothing could then prevent the likely evolution (gradual transformation) of the signifier's acoustic form to match the expression to the apparent features of the concept expressed. "Under those conditions, the arbitrariness of the sign would be sacrificed on the altar of expressivity."[61] In actual languages, however, the autonomy of signs is ensured by the fact that this "slipping off" of the signifiers is prevented by their articulation into phonemes (one-dimensional units) to which the meanings of words have no access. The different phonemes exclusively defined by intralinguistic relationships are in solidarity with the occurrences of allophones and thereby strongly resist modification.[62]

In summary, the fundamental significance of the double articulation is that it causes the form of the sign to remain a two-sided form. That signs are defined by distinctions that are decoupled from meaning, that is, from the form's outside, is precisely what ensures their isolation and at the same time constitutes the condition for redundancy necessary for reproducing signs. This double articulation is found only in language: Other "sign systems" are arbitrary yet not articulated (as in the case of a traffic light), or they are articulated yet not arbitrary (as with ideogrammatic script or pictorial communication in general). Only a doubly articulated system has an absolutely differential nature.[63] In all other cases "positive" elements are involved. That is why language is the condition allowing other types of sign systems to exist. Without the isolation ensured by language, the autonomous production of signs would be impossible.

It remains to be investigated how the network of distinctions of the indication/distinction type succeeds in defining linguistic elements such that identity and distinctiveness, iterability and productivity can be combined.

V

Earlier we credited the network of i/d distinctions with the function of producing sufficient redundancy for the connectivity of operations. Redundancy means that it is not necessary to observe an element directly to gain the information it imparts: Knowing other elements already reduces its surprise value.[64] But how does that work in the case of language? How can signs be specified in each use, despite the generality they need in order to apply to many different contexts, such that they effectively limit the number of possible connections? How is it possible that a linguistic sign both stays the same—that is, remains recognizable—and is always different and therefore can be made to fit the context?

Let us return to Saussure once again. Without transcending the plane of language-internal i/d distinctions,[65] he discerns two types of relations between linguistic elements:[66]

> *Syntagmatic relations* (*in praesentia*) between words appearing in the same sentence (or syntagma). In this case, the value of each element depends on oppositions to the preceding and succeeding elements in the chain of signifiers.
>
> *Paradigmatic relations*[67] (*in absentia*) between words that "have something in common" and therefore refer to one another. For example, "en-

seignement" refers to "enseigner," to "armement," or to "éducation" in a principally infinite network.[68] Here the value of each element depends on opposition to alternative possibilities.

The "intersection" of the syntagmatic and paradigmatic "axes" in every single element yields the special combination of determinate and indeterminate that is the basis of the productivity of language.[69] Each communication is new and distinct from every other; that is why each word employed is uniquely specified depending upon its syntagmatic relation to other words in each sentence. Linguistic elements are sufficiently general that they can take on meanings relative to each context. Yet at the same time they remain instances (*tokens*) of a unit (*type*) that is defined by the paradigmatic relations to other linguistic units. In this sense they remain indeterminate (and recognizable) with regard to their contextual employment: Every word can acquire many more meanings than the one corresponding to it in a particular sentence, and every word also points exactly to this surplus. In the parlance of observation theory: What is signified in a present observation is determined in contrast to all other operations of the same system—thus always self-referentially; this happens simultaneously for both the "adjoining" observations and those observations "condensed" within the structure of the system. Moreover, this happens without involving the s/e distinction: The syntagmatic and paradigmatic axes are only concerned with intralinguistic distinctions.

The crossing of the two axes serves to determine both the monemes and the phonemes. Both the units of the first articulation and those of the second are defined in an unambiguous manner only by negations, that is, without reference to any positive content, and those negations occur on two different axes simultaneously. According to Barthes, dividing the "text without end"[70] implied by every syntagma constitutes the first problem of signification; inserting distinctions is hence the first feature of observation. The word "bank" in the sentence "No camping on the bank" acquires a specific meaning only in contrast to the other words of a syntagma; this makes it possible, among other things, to exclude the meaning "financial institution." However, signification only happens if the meaning of the word "bank"—that is, its position on the paradigmatic axis—is known. The selection resulting from the selection on both axes makes it possible to isolate a signifier in such a way that a distinct meaning corresponds to it. The same principle also applies to the second articulation: The phoneme "p" is recognized in the enormous variety of contexts in which it occurs[71] as the same since in every instance it is sufficiently determined by the differences from neighboring and alternative sounds. That is why "punch" is not confused with "bunch," for example. Once again, the issue here is the production of redundancy, which is also proven by the fact that one can usually reconstruct a missing sound or word.

The simultaneous placement on two axes shows concretely how language's "negative units" are constituted; it also reveals how sufficient specificity can be obtained on the basis of relations that are exclusively internal to the system. The "outside world" as such is never involved. The "segmentation" of the signifier is achieved solely through intralinguistic i/d relations. Here the external reference of signs comes into play; its role, however, is limited to "checking" the correctness of the segmentation, and the external reference does not intervene in the operational determination of form. Although there is recourse to meaning, it remains purely formal: there is no meaning as such based on any kind of "substance," but only meaning as an index of the signifier; that is, meaning "*places* the signifier, that is all."[72] This is demonstrated by the well-known "commutation test,"[73] which serves to check the correctness of the division of the syntagma into morphemes and phonemes by substituting one part of the signifier for another expression and by observing whether this alternation is accompanied by a corresponding change in the signified. If this is so, one speaks of autonomous articulatory units; otherwise one is dealing with "variants" of one and the same unit. To identify a phoneme, for example, one examines whether it has a distinctive function: the commutation of one phoneme for another can reveal a difference in content (for instance, "punch" vs. "bunch"), or it can reveal that a corresponding meaning does not exist (for example, "punch" vs. "sunch"). This does not occur when one variant is exchanged for another of the same phoneme (for example, two different pronunciations of the "a" in "tomato"): One still does not speak of two autonomous units even if they are further apart than the phonemes "ā" and "ä."

VI

The double articulation of language enables us to highlight those mechanisms that establish the functionality of language as a medium for constituting forms on other levels. Now the guiding distinction is the *distinction between form and medium* that Fritz Heider put forth and Luhmann generalized.[74] Here, the distinction of the two sides concerns the coupling of elements. The elements of the medium are loosely coupled, while the elements of form display tight couplings, which allow the forms to impress themselves into the flexible medium just as a footprint imprints itself into the sand. For the distinction between form and medium to be realized positively, the elements must satisfy some requirements—and those requirements are to be found most clearly in the elements of language.

In certain respects, the medium as such must not exhibit forms of its own, nor should it become an object of observation itself.[75] The second articulation into meaning-free phonemes matches this availability of an order of units that are not forms themselves (they are not signs). And usually one is also not aware of the division of the acoustic continuum that is the basis for the constitution of linguistic forms—nor of the signifying relationship as such.[76]

Moreover, the medium must contain more possibilities than are actually used in the constitution of specific forms, which always presuppose a selection.[77] Language fulfills this requirement by segmenting the phonetic continuum and thereby producing first the morphemes and then the phonemes. The second articulation ignores other possible sounds (no language actualizes all acoustic possibilities), and the first articulation disregards other possible sound combinations. In a particular language there are always more sound combinations possible than those actually employed by the words of that language. Italian, for example, does not realize combinations such as "lipa" or "malare." In other words, none of the forms is necessary. Forms are inescapably contingent in the sense that they could turn out differently if the selection were to proceed otherwise. The words of a language are indeed arbitrary: they are contingent upon a selection that could also have turned out differently. If this were not true, and if the signifier/signified correlation were dictated by the lack of other possibilities (i.e., were necessary), then obviously the isolation of language could not be guaranteed.

Medium and form, incidentally, are correlative notions only existing as a pair: There is no medium without forms, and there are no forms without a corresponding medium.[78] But how do we conceive of the elements?

Do they or do they not precede the constitution of forms, which also simultaneously constitute the medium? In other words, are elements that are to be coupled independent of the distinction? In the case of language, the combination of the s/e and i/d distinctions shows how the elements of a medium can be at once both dependent upon and independent of forms. To identify the phonemes of a language, the commutation test is used, which in turn refers to meaning (thus to the sign realizing the s/S distinction). Yet the identity of each phoneme is exclusively defined by the intersection of the i/d and s/e distinctions on the syntagmatic and paradigmatic axes; their identity is hence independent of the words in which they appear. As mentioned earlier, the phoneme "t" can occur in many different words, yet does not depend on any one in particular. If, however, it did not contribute to realizing forms as s/S distinctions, it could not be identified.

Finally, the distinction between medium and form has the ability to repeat itself and thereby to produce increasingly nested forms:[79] Forms constituted on one level can serve as a medium for creating more forms. Language's double articulation demonstrates how one and the same mechanism (i.e., determining a unit through the intersection of the syntagmatic and paradigmatic axes) can be applied on several levels in an increasingly nested manner. The phonemes impress themselves as forms into the acoustic medium. They then serve as a medium into which the morphemes' forms imprint themselves. The possible combinations of words are finally the medium for the form constitution of sentences. To be sure, this is not a case of simple repetition: The "economical relation" between monemes and phonemes has shown that the constitution of forms, when moving up to a higher plane, performs a selection on a selection which, starting from a smaller number of units on the lower level, allows the specification of an ever-increasing number of units on higher levels.[80] With scarcely more than twenty phonemes, several thousands of words can be generated, which can be combined into infinitely many different sentences. On a higher plane, forms combine increasing selectivity with expanding freedom—and language shows how this happens.

Yet the most important contribution the form analysis of language has to offer for understanding the distinction between medium and form concerns the role of language in the interpenetration of psychic and social systems. If we assume that both psychic and social systems are in fact operationally closed, autopoietic systems, then we have the problem of explaining how their "coordination" is ensured, given that the socialization of every consciousness and the very possibility of communication rests on this coordination. Any communication requires participation of at least two psychic systems, which, however, always remain in the environment of the communicating social system. How can the separation of the two levels constituting systems be reconciled with the fact that in any communication event a social operation coincides with a psychic operation? Niklas Luhmann attributes this coordinating function to language:[81] By virtue of its "formal conciseness," language enables the two types of systems to use each other as media and to adopt the imprinted forms. It is thus expected that all psychic content can be expressed linguistically, and likewise that all communication can potentially "fascinate" consciousness by way of language. Analyzing the form of language and, especially, investigating the double articulation permit one to determine the concept of "formal conciseness" more precisely and to explore the fundamental mechanisms involved. Language's capacity is connected to the fact that language is able to build up forms without necessarily being confined to meaning. As shown, the separation of the two i/d and s/e distinctions leads, among other things, to the fact that the determination of linguistic elements proceeds relatively independently of signification. A word is at first determined in relation to other linguistic elements and not with respect to a referent. For communication that uses language this means that the coordination of the systems can take place simply on the level of the "division of the signifier" without the inclusion of meaning. That is to say, coordination is ensured independently of whether or not real "understanding" is reached. Communication can continue even when words are interpreted differently. Eventually, the processes can coincide in one linguistic operation without a necessarily identical constitution of meaning. In Luis Prieto's words, "simply using the code does not allow a user to compare his way of analyzing the existing classes of sememes with that of others"[82]—and it is also not required.

DIRK BAECKER | **THE FORM GAME**

> Only two can play this game.
> James Keys [alias George Spencer-Brown], *Only Two Can Play This Game*

The combination of elegance and gracefulness (in Italo Calvino's sense)[1] is one of the most intriguing features of George Spencer-Brown's calculus of form:[2] What at first appears to be determinate turns, upon closer view, into something indeterminate, which when viewed even more closely turns out to be a receptivity for more precise distinctions—a receptivity that, instead of bringing the process of determining to a close, passes it on to the next observer. This kind of indeterminacy affects the "mark of distinction" the moment the mark is "re-entered" into the domain of what is distinguished. This "re-entry" of the distinction into what is distinguished alters the status of the distinction, without revealing precisely how that happens. Spencer-Brown remarks only that the re-entered distinction should be taken as "non-literal" (p. 65), that is, not as an injunction to draw a boundary and indicate the inside of the distinction in contrast to its outside. The re-entered distinction does not have the same properties as the injunction "Draw a distinction." Rather, it represents itself on the inside of the distinction into which it is re-entered. This representation deprives the injunction of the very instructiveness that the injunction shows at the initial distinction. Practice turns into a recipe that one may or may not follow. Spencer-Brown therefore distinguishes between "cross" (meaning the distinction as an operation) and "marker" (signifying a re-entered distinction): "Thus a cross is a marker, but a marker need not be a cross" (p. 65). Every distinction can be indicated, but not every indication makes a distinction.

The difference between a distinction and a re-entered distinction is that only the re-entered distinction can be observed on both of its sides (that is, it can be observed in its form) within the domain of what it distinguishes. That is why Spencer-Brown says the re-entered distinction "represents" the boundary of the space into which it is re-entered. The distinction becomes autological and thus paradoxical: it is no longer what it is. But only by virtue of this paradox does the distinction become noticeable. We have to accept the paradox if we want to move from constructing the form to knowing the form (p. 6), that is, from the injunction to make a distinction to the description of the distinction made (p. 77). We might even deal with the name of a form that is copied into the form. This name is, as all names are, distinct from the form (p. 6). We do not know exactly what we are dealing with since in order to know, one must first know one's position within the arrangement of the form, then calculate according to the laws of form whether one is dealing with plain or re-entered distinctions. The problem with the re-entered distinction, however, only occurs when one is dealing with second-degree equations, that is, with nonfinite expressions and hence also with indefinite solutions. But is it precisely here that we lose contact with arithmetic since, as Spencer-Brown states, the "excursion to infinity undertaken to produce it has denied us our former access to a complete knowledge of where we are in the form" (p. 58). Object-related observations as well as what they indicate only occur insofar as they occur as *eigen*-values, that is, as recursively repeated values of second-order observations.[3]

But as soon as we do not know any longer where we are within the form, we cannot tell whether any given distinction is a plain distinction or a re-entered one, that is, a "cross" or "marker." We have a choice, we can make a choice, we can commit ourselves. So that there will be no doubt about the systematic and endemic nature of this indeterminate state, Spencer-Brown instructs us to imagine the baseline of a form arrangement, that is, any level on which "first" distinctions are made, as being framed by an "unwritten cross." That is to say, every construction of a form always takes place within a constructed form. Pointing out this connection between the "cross"/"marker" distinction on the one hand and the instruction regarding the "unwritten cross" on the other, Matthias Varga von Kibéd concluded that the next step would be a calculus capable of considering implicit contexts in a formal way, or more precisely, a calculus that achieves explicit "formalization of iterated thematizations of implicit contexts."[4] This task, however, must fail to the extent that the world itself becomes the only possible solution. Any case of explicit formalization moves the outside of a distinction to the inside of a distinction, which in turn implies an outside.

Spencer-Brown's calculus of form can be regarded as an arrangement of re-entered distinctions. He advances the formalization up to the point that any formalism can appear as a re-entry of forms. Taking the formalizing procedure any further would lead to an infinite regress, that is, to a recursive equation whose *eigen*-value is the calculus itself. The merit of the calculus lies in its operation of re-entry as well as in making this operation visible as an instruction to construct. The achievement here consists in reinserting the operation of re-entry into the calculus, then clarifying that any distinction can be observed either as a "cross" or as a "marker" and hence that we may suspect the "unwritten cross" of being an "unwritten marker." The maneuver of the calculus of form could thus very well be compared to Buddhist meditation practices or Don Juan's maxim "stop the world":[5] in both cases, reflecting emptiness, that is the "unmarked state," reveals the abundance of forms, though transformed, toned down, and dramatized. A "marker" can emerge where a "cross" used to be, but only at a price: not only the "marked state" but also the "unmarked state" (or the "être de non-étant"[6] as Lacan formulates it with reference to Freud's death drive) must be re-entered into the domain of what is distinguished.

A sociologist can hardly imagine that mathematicians, Buddhists, and American Indian shamans have discovered a phenomenon that would not correspond to any social practices. The emptiness and abundance of forms can be experienced socially as much as mathematically and meditatively. As soon as they are communicated, both the mathematical and the meditative procedures are nothing but social practices—albeit social practices acquired at a distance from the social, that is, by virtue of a "first" distinction that can be regarded as re-enterable in different degrees. We may assume that for different societies the odds that mathematical, meditative, or social distinctions will be re-entered are different. Controls on religious practices can be lifted in societies in which the social order is not at issue. Conversely, a society could easily increase in reflexivity while committing mathematics to rigidly logical forms. Yet before we can envision a way to prove such sweeping conjectures, we need to show that what the calculus of form demonstrates for mathematics and logic and what Buddhist meditation and American Indian lore make accessible to human consciousness is also something conceivable in the realm of social possibilities. The calculus of form, meditation, and lore would presumably not be what they are if they had not already been informed by the possibilities of the social.

There is a social practice that seems to correspond to what Spencer-Brown investigates under the heading of "re-entry": play. We can even go a step further: play is presumably not just one social practice among others but the social practice that has to be presupposed as a possibility in all other social practices. Other social practices come about by eliminating certain properties of play. Play is the social phenomenon *per se*. It is true that playing is not restricted to the social domain in a narrow sense. But when we speak of the play of atomic particles, the play of wind and waves, mind games, or the games of the gods, a specific form of socialness can be noticed even though there is no society involved.

In accord with Talcott Parsons, we define socialness as the double contingency of expectations attributed to *alter* and *ego*.[7] By "practice" we mean any coupling of operations such that the exclusion of certain actions enables the inclusion of other actions. Practice thus is geared toward a selectivity that obtains its meaning from both what is avoided and what is done. If one accepts those two definitions as well as the possibility of combining them in the notion of social practice, then one confronts a paradox, namely the paradox of the selectivity of double contingency: the paradox consists in the fact that a situation cannot be both selectively determined and doubly contingent. Following Niklas Luhmann, we can assume that any selection, however it may come about, converts the double contingency into a single contingency so that communication can be connected to communication and action to action, and a social system evolves.[8] Hence, either situations are doubly contingent because they lack any selection that could serve as a reference point for determination, or they are already structured by a certain selectivity, in which case a single contingency overrides the double contingency of complementary expectations.

The purpose of construing social practice in the specified sense and at the same time letting this definition run aground on a paradox is to underscore that the problem of double contingency is not at all solved, but rather utilized, by the emergence of a social system. The problem remains; it triggers and motivates a solution that contributes to the determination of a social situation, but it does not disappear. In fact, at any moment, the problem can reappear. The paradox informs us that social situations derive from the impossibility of narrowing them down to the specifications upon which they rest. In every social practice the emptiness out of which it grew can be detected. In the eye of the observer every distinction has a form. Any social practice can be observed with respect to alternatives that the contingency of its selections brings into view.

Furthermore, the notion of social practice is Janus-faced. On the one side, practice is nothing but action in a certain concatenation. On the other side, practice is all communication, through which alone the concatenation comes about; communication that attributes each selection to actions and that recursively figures out the possibility of further actions from the selectivity of every single action. The selectivity of actions does not emerge by itself, but rather is the product of attributions communicated accordingly. An element of recursive construction is thereby introduced into practice since these attributions relate backward and forward to other attributions whose contingency underscores the contingency of the present attribution. Communication is the medium of social practice, for it is only within communication that the problem of self-transcendence can be solved by way of exclusively immanent operations.[9] In every social practice communication is the "unwritten cross."

According to these conceptualizations, to call play social practice *per se* means to describe it as a concatenation of operations. Carrying out those operations informs us that every such concatenation is available in social situations characterized by the implied double contingency of complementary expectations. In play, socialness is constituted by way of reflection onto itself as the other side of itself. In play, socialness is experienced as what it is, namely as contingent, roughly meaning that it is neither necessary nor impossible, or again, given yet changeable.[10] Play in general reveals the form of the social by which the play infects the world.

It is not new to conjecture that something happens in play that, owing its existence to play, can only be grounded in play.[11] This is already explicit in Wittgenstein's idea of language game: "We can . . . think of the whole process of using words [in a context of pointing and naming] as one of those games by means of which children learn their native language," says Wittgenstein.[12] One must already have mastered the game to be able to learn a language, he states.[13] But obviously the converse is true as well: Without first having learned a language, one cannot play.

In mathematical-economic game theories, too, play takes on a constitutive character. After all, it is meant to describe those conditions that lead to interdependency despite the independence of the actors.[14] The game itself, however, is described by the totality of its rules.[15] And, mind you, there is no playing with the rules, as there is no playing with the game itself. Play in game theory is not meant to make a context available, as in Wittgenstein's and other theories of play, but to establish beforehand a context for profit-maximizing intentions that subsequently cannot be reconsidered—intentions that, even when one's information is incomplete or when one is playing against irrational players, warrant the very rationality on which the coordination in those kinds of games rests. If one is to imagine, within the context of game theory, playing with the play, one has to imagine a non-zero-sum game in which the fictitious player $n+1$ is the game itself—a game that can result from the sum total of the moves of all n players, winners and losers alike. One would then see that with their play, the n players put their game on the line.

Again the conjecture is not new; likewise a corresponding concept of play is well established. New, however, is the possibility of substantiating this conjecture by means of a calculus that reveals how something can be constructed out of nothing but itself and a world of chances, a calculus that therefore lets us reflect on the conditions for construction. I have limited what follows to an examination of this conjecture using preexisting considerations; the elaboration of a sociology of play I shall leave to future research.

Calling play social practice *per se* in the aforementioned sense amounts to a language game that clarifies its own conditions to the point of understanding three steps of Spencer-Brown's calculus of form, namely those steps made visible by the re-entry of the distinction into the domain of what is distinguished: (1) differentiating between "cross" and "marker"; (2) discovering (thus marking) the "unwritten cross"; and (3) describing the form of the distinction as the unity of "marked state" and "unmarked state." This clarification arises from the discovery of the paradox that unity (for instance, the unity of distinction) is not what it is. This paradox is unlocked by difference and is thereby transformed into observation; it is an element of the "deliberate complication" Stephen Miller described as a property of human and animal play alike.[16] Matters become more complex and no longer are what they were; yet they become observable. Play introduces obstacles and complications that turn out to be precisely those distinctions whose form is to be explored. Eventually we discover that we have been dealing with complications all along which conceal nothing but the fact that there is nothing to complicate.

Gregory Bateson has already developed the very concept of play we need in order to understand these three steps. For him as well, play is a practice, which is to say, a concatenation of operations. There is no play that can be reduced to a single move. Bateson captures this fact with the formulation that "'play' is not the name for an action; it is the name for a *frame* of an action."[17] It is a frame that enables actions; in the course of those actions the frame becomes visible and explorable as the constitutive condition for all actions.[18] The key to this analysis is provided by Bateson's discovery of a paradox of play matching the paradox of the re-entered distinction that is not what it is. The paradox of play is that one can perform actions in play that do not mean what they mean.[19] The bite of a playing dog is actually a pinch and not a bite, yet it stands for both the pinch and the bite. Play simulates what it dissimulates. And the first and most important demand put upon the players is to make and comprehend precisely this distinction between the dissimulating simulation and what it signifies.[20] "As long as we are only playing we do not cross a Rubicon—neither in war nor in love."[21] Translation: In play we are dealing with "markers" and not with "crosses."

Yet the players can draw the distinction between "crosses" and "markers" only if the "metacommunication" "this is a play"[22] is part of the game. To be sensitive to this metacommunication is thus the second demand placed upon players. "This is play" is the very distinction that singles out play from everything else and permits us to establish play as a free, self-contained, and comparatively unusual form, as Johan Huizinga discussed.[23] This distinction has to be presupposed explicitly as a distinction implied in all subsequent moves. The distinction is the "unwritten cross," no longer appearing in the play itself once it is in motion; making the "unwritten cross" explicit inevitably spoils play. For the most part it is the rules of the game that have to perform the always precarious and problematic task of implicitly making the explicit distinction "this is a game" by explicitly allowing certain ways to concatenate moves and prohibiting other ways.

Furthermore, contemplating the rules makes noticeable the form of play as a re-entry of the distinction into what is distinguished. The rules have an inside and an outside. Their inside is the designation of the game whose possible moves are mapped out by the rules. Singling out a game from everything else is their outside. However, no rule would work without bringing to bear on the inside of the game the distinction "this is play" as one belonging to the outside from which the game is singled out by means of this very distinction. The rule therefore repeats the paradox of play within play itself: The outside of the rule, that is, the exclusion, becomes the inside. Play becomes the play of the world. And the inside of the rule, that is, the game's designation, becomes the outside. The world turns into play.

And yet there can be no mistaking the world for play and vice versa since play—by reflecting onto itself, that is, by re-entering into the play the frame that marks the game's outside and inside—discovers and describes itself as social practice. This social practice, geared toward a certain concatenation of operations, takes this concatenation as a sign of the selectivity of its own practice knowing that it has, along with the constitution of this selectivity, another side. Unfolding itself as a social practice and thereby keeping open the possibility of inserting the reference to the other side into the double contingency of complementary expectations attributed to *ego* and *alter ego* is one of the moves of the game. Anyone who wants to play the game—and there is no other to play—must be involved in this minimal selectivity, which is at the same time the reason that the player must reckon with seeing through the game and discovering its other side.

MICHAEL HUTTER | **THE EARLY FORM OF MONEY**

Proto-Forms

We are constantly involved in monetary transactions, yet we hardly ever notice the media we use in such transactions. And we should not notice them, either, since otherwise our trust in their independent functioning would be undermined. Observing the early period in the evolution of the monetary medium allows us at least to obtain a certain distance from our own communication environment. In observing the preserved coin forms of the ancient monetary medium we gain a sense of how arduously and slowly communication regarding economic values spread and solidified. Material forms of money have the advantage that they can be easily, that is, visually, observed. Whenever we speak of the "coin form" or "the form of the coin" hereafter, this notion is meant to encompass three different distinctions: First, it signifies the quite obvious delimitation of a form from its material environment. In this environment the form becomes cognitively, that is, psychologically and physically, observable for human eyes and hands. Second, the term refers to what is signified by a monetary sign, that is, value. When coins are used, this value is at once generated and passed on. And third, the "form of the coin" denotes the communication of payment that is structured by this form, because only in the structured form does the medium become observable.[1]

Numismatists and archeologists concur that in the course of the seventh century B.C. the Mediterranean civilization started to use precious-metal coins to make payments. By using available archeological and literary sources, we will follow this development until the emergence of the Athenian "owl coin" in approximately 500 B.C. In order to recognize the changes in form more readily, we shall divide our account of this process into a sequence of five basic forms.

During the 500 years prior to the emergence of coin objects, bean-

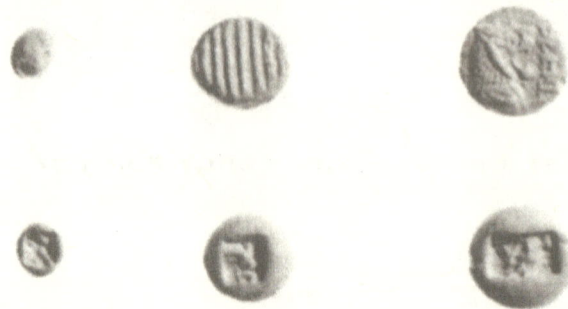

From left to right:

Form 1: Ionia, electrum, $^1/_{48}$ stater, ca. 650 B.C.
Form 2: Ionia, electrum, $^1/_4$ stater, ca. 650 B.C.
Form 3: Lydia, electrum, $^1/_6$ stater, ca. 600 B.C.

From left to right:

Form 4: Aegina, silver, didrachm, 550 B.C.
Form 5: Athens, silver, tetradrachm, 520 B.C.

Source: Colin M. Kraay, *Archaic and Classical Greek Coins* (London, 1976), figs. 50, 51, 63, 114, 175.

shaped nuggets made of silver, electrum, and gold were in use in the Eastern Mediterranean region.[2] The production technique consisted in melting metal pieces of the desired weight and then flattening them. In their sizes, these so-called *phthoides* corresponded quite accurately to the units of weight employed in each of the local measuring systems.

The discovered *phthoides* have weights ranging from 20 grams down to 0.13 grams. Even the smaller pieces had enormous purchasing power. One could use them to buy a goat, a slave, or the yearly services of a mercenary.[3] Larger pieces were used exclusively in long-distance trade. For example, among the dozens of value-measuring instruments used in Mesopotamian societies, larger *phthoides* played a significant role in balancing the values of goods agreed upon in the exchange.[4] This practice of payment remained stable until about 700 B.C. Subsequently, punchmarks appeared on some nuggets. (See form 1 in the illustrations.)

The punchmarks are signs—but signs of what? The discussion about the meaning of the marks has a long and controversial history. The most prevalent opinion today among numismatists is summarized in the following quotation: "The merchant's mark was often no more than the end of a particular broken iron nail hammered into the metal, but it could be readily identified by the man whose mark it was. This eliminated the necessity for the weighing and testing for purity of each piece of the precious metal every time it passed through the merchant's hands; he knew his mark."[5]

The above quotation highlights the significance of *recognition*. The mark communicates a quality of the materials thus signified. This is achieved by a combination of two factors: the physical change the hammering left behind and the similarity to signs already in use. The continuity of the assessment of this quality is assured by the limitation of the assessment to a single dimension within which person and locale remain the same. Only in the memory of the evaluator does a comparison of his observations of the objects at two points in time occur. In this way signs can emerge: as perceptions within memory, as signs for one's own observation. The "sign in and of itself," however, can also be noticed by other observers, namely as a *Versprechen* (promise; slip of the tongue) of validity transcending the present moment.[6] The one who marked the metal piece is inside. One can attribute to him a *Versprechen* if one is positioned within the engraver's environment.

In this context of gradually evolving possibilities for expressing recognition and *Versprechen*, payments in the engraver's household and his immediate surroundings could occur faster and more frequently than they could with unmarked metal nuggets.[7]

In assuming that the "engraver's household" was the relevant form of agency at that time, I have in mind the organizational form of the *oikos*, that is, a community of several dozens or even hundreds of people living together. To what extent did payments play a role within archaic royal and merchant households? Internal relationships were familial or political. Yet on the household's periphery, exchange was unavoidable. The *oikos* was surrounded by workers and soldiers offering their services in exchange for food and money, and by merchants who supplied goods not produced by the *oikos* itself. At this periphery, marked *phthoides* started to circulate faster than the unmarked pieces.

The Selection of the Form

The next form appeared sometime around 670 B.C.[8] and can only be found on electrum pieces from the region between Sardis and the Ionian coast of Asia Minor (see form 2 in the illustrations). These coins are named after their basic form, *quadratum incusum*. They are flattened, and on their reverse one or several markings are hammered, which have obviously lost their authenticating function. The markings are independent of each other, yet generated in one stroke.[9] The side that rested on the minting punch often shows irregular imprints caused by the anvil. On some pieces found at Sardis, the obverse shows regular striation (see form 2 in the illustrations). There might be a technical reason for the striation: it might have helped prevent the metal flan from slipping off the punch. Yet as in the case of the punchmarks, this structure, too, took on a communicative interpretation: the marks occupied the reverse of the piece while the striation covered the obverse.[10] Thereby the entire piece became perceivable as a structured form. This is the form that came to be called "coin." We must now clarify why the new form only appeared in electrum pieces and only in a small region around Sardis.

Sardis was the city where the trade route from Assyria to Greece branched off to the Ionian cities.[11] Sardis was thus located at the intersection of two completely different cultures. In the east, the Mesopotamian state-societies continued to exist after the Assyrian conquest of 745 B.C. In the west, the migration of Ionians led to a chain of small settlements along the coast. The basic institutional structure of these settlements still resembled those of their rural ancestors. Accordingly, the communication of payment was different in each society. In Assyria gold and silver ingots, in addition to other materials, were used as media for the storage of value and for the balancing of exchange values. Moreover, gold and silver were the subject of detailed metaphysical discussions in which gold especially was granted a key role based on an analogy with the sun deity.[12]

The degree of the material's purity was kept constant, and the weight was measured with technically elaborate precision. As long as the parity between the value-measuring instruments remained unchanged, this method had the advantage that it could measure economic values consistently up to large value accumulations. However, the complicated measurement procedure restricted the use of these media (such as gold and silver ingots) to a small group of merchants, temples, and royal houses. The Assyrian social system did not have a generally accepted and understood monetary medium. Its fundamental guiding distinctions were still determined by deity and power.

Among the Ionians the situation was significantly simpler, and accordingly the requirements for the media of communication of payment were much less demanding. As in other primitive cultures, here cattle and certain implements were employed as means of payment. The implements always had a religious connotation. Spits, trivets, axes, and rings were part of sacrificial rituals in which they were "handed over" to the deity. What worked in relationships to supernatural powers was also considered trustworthy enough for exchanges between people. In addition, the implements were magically protected against unauthorized alteration by dint of their religious connotation.[13] The materials employed for the implements ranged from iron to copper and bronze, and even silver; gold was hardly ever used. Money in the form of implements had the advantage of being composed of single, discrete elements that were easily countable. The valuation, however, remained limited to the narrow confines of a particular tribe or a settlement. Those objects remained worthless for making tributes or trading in distant locales.

The striated electrum coins from Sardis could be used as a monetary medium in both cultures because the pieces held two different meanings; they were interpreted as signified metal (with a shine of gold) in the east, and as metallic signs (with a shine of silver) in the west. The effect was quite literally unconscious. Only electrum, found naturally in the fluvial sand in the Sardis region, carried the meanings of both gold and silver. Moreover, only the closed form of the marked coin could at once be interpreted as a credit-worthy sign and as a credit-worthy object. The improbability that such a double ambiguity would occur is obvious.[14] As far as we know, this event did not recur in any other place at any other time.[15]

How did this new form of coin from Sardis change the credit-worthiness of payments? In several ways: The shape of the markings not only referred to the one who marked the coins but also related to the magic protection invoked by the act of sealing: One is not permitted to "open" this form. The form's closure made alteration technically more difficult since the regularity of the pieces made them more recognizable. But above all it was the surplus of meaning resulting from the multiple interpretations of the new form that led to an immense expansion of the communication of payment. The monetary units were used by more economic participants in increasing numbers of transactions, and accordingly they circulated faster. Consequently, the money supply that issuers of the novel coinage had at their disposal increased.

If the resulting growth in buying power and financial strength was in fact so tremendous, then surely we should expect to find traces of it in historical accounts. And indeed this is the case. There are clear signs that in Lydia around 670 B.C. the financial strength of merchants rose so greatly that they were able to assume political power and thereby establish a new form of political rule: the *tyrannis*.[16]

Literary sources mention for the first time a "merchant king" by the name of Ardys during the period 766 to 730 B.C. He is said to have ousted the prior king with money he accumulated as a "landlord and coach builder." Further power struggles ensued between feudal aristocracy and merchants who could suddenly afford mercenary troops. Finally, with Gyges reigning from 687 to 652 B.C., a form of rulership took hold that based its claim to power not on birthright but on money.[17] This astounding change in the balance of power within Lydian society thus occurred in the same time period in which the striated electrum coins first appeared. The significance that theory has assigned to this coin form can thus be confirmed historically.

The Medium of Money and the Form of Power

Gyges founded a new dynasty, the Mermnadae, which lasted until the defeat of the king Croesus by the Persians in 546 B.C. The consolidation of Croesus's power and his legendary wealth, we can presume, resulted from Gyges' turning the minting right into a state monopoly.[18] This step was associated with a new form (see form 3 in the illustrations): instead of the striation, an image of a lion's head[19] was struck into the electrum flan; the reverse with its square marks, however, remained unchanged. The lion was the totem animal of Astarte, the highest Lydian deity. This image transferred a mighty, religiously coded message to the coinage.[20] The coining merchant who had seized public power was now able to use public images. Originating in religious discourse, the communicative form of the totem animal initially denoted the political system and subsequently served to secure the credibility of a given economic value. Credit-worthiness was thereby increased in several ways: all members of the political-religious community were familiar with the image; material alteration of the coinage took on high magical risk; and at the same time the royal household guaranteed that the circulating coins would be taken back.

It is not surprising that Gyges monopolized his discovery that political power multiplies miraculously from the purchasing power of coins. Such a restriction of competition, however, could only be established within one's own sphere of influence. The cities within the domain of the Ionian settlements quickly imitated the new form—an electrum flan with totem animal (or totem body part) on the obverse, ornamental punchmarks on the reverse.[21] The first cities to do so were Miletus, Ephesus, Samos, and Phocaea, followed in a second wave by Smyrna, Chios, Cyzicus, and Lampsacus.[22] The minting right was usually held by the city government, yet private issues of coins also occurred.[23] Long-distance trading of standardized goods, intensified regional trade, and investments in agriculture and crafts became part of the economy via the new coins, meaning that they became the subject matter of the communication of payment.

It is important at this point in our historical reconstruction to emphasize the slow pace at which the described changes of the electrum coinage took place. Roughly half a century lay between the emergence of the striated coins and the coins minted by the Ionian coastal cities.[24] The circulation radius was still small, and the relatively few, high-quality electrum pieces still constituted only a fraction of the economic transactions. A clear initial stabilization of the form had nevertheless taken place during this long period of time. By means of improbable ambiguities, and originally in closed trade circles, variations were infused into the code of payment. These variations changed the code and subsequently its phenotypical environment. And this process did not require a conscious decision on the part of its economic players.[25]

Stabilization of the Form

Decades after their introduction, the silver content of the Ionian coins had risen to almost a hundred percent. They were nevertheless traded nominally at the price of electrum.[26] The first definitely silver currency was minted around the year 600 B.C. on Aegina.[27] Since the Aeginetans were a clan of long-distance traders active in the eastern Mediterranean, it is likely that here, too, an ambiguous development occurred within a closed circle. The new coinage initially was by no means accepted among the various trading partners of the Aeginetans; the clan used it internally for exchange and storage of value, that is, members of the clan communicated with each other by means of the silver coinage.[28]

The pictorial form of the new coinage remained constant: on the obverse a turtle is imprinted—the heraldic animal of the Aeginetan Hera—and on the reverse there is a *quadratum incusum* that had finally become simply ornamental. The change to the next form (see form 4 in the illustrations) was limited to one single dimension, namely that of the coin's material. The lower material value of silver made it possible to mint fractional denominations. Therefore, for the first time, local trade became accessible to the new form of payment as well. Moreover, a weight system was employed for the first time in Aegina which was tailored to the peculiarities of the monetary code, especially to the value ratio between silver and gold.[29] The artificially created communicative qualities of the monetary medium had become more important than the naturally existing properties of the material itself. The material used, incidentally, was no longer a local natural resource. Silver had to be imported from outside of Aegina before it could be minted.

At the beginning of the sixth century B.C., silver currencies with different systems of measurement spread across many Greek cities. But only around 520 B.C. did the general form emerge that would remain unchanged until modern times. The images once again shifted. The animal image moved from the obverse to the reverse. The punchmark, having been part of the coin sign for a century, degenerated into an insignificant ornament and subsequently disappeared. The obverse was now imprinted with a human head surrounded by the attributes of a Greek deity (see form 5 in the illustrations).[30]

Taking the example of the "owl coin," we can once again observe how slowly the form of the coin changed.[31] The first Athenian coins displayed about a dozen different obverse types. These types have usually been interpreted as heraldic symbols of the coin-minting families; for that reason they are called *Wappenmünzen* (heraldic coins). All variants, however, followed the Attic coin standard and exhibited the same, almost identical, marking on the reverse that identifies them clearly as Athenian coins. The coins were evidently used in the immediate vicinity of Athens for local and regional commodities. The sign on the obverse had become arbitrary and hence variable; the mark that guaranteed the weight was still part of the archaic-abstract sign of the reverse. This, however, allowed only for internal communication since only those who were familiar with the abstract sign could decipher it.

In the years around 520 B.C., an animal, initially a lion and then a bull's head, appeared for the first time *within* the incuse square on the reverse of an Athenian coin. Shortly thereafter, the tyrant Hias, son of the tyrant Pisistratus, "reformed" the Athenian currency by replacing the circulating coin variants with a single form.[32] At this point the reverse exhibits, still within the incuse square, a standing owl with an olive branch. The symbol of the owl, the totem animal of Athena, signified Athens for external observers as well.[33] The obverse showed Athena's head with a helmet. Athena was more than the patron deity of the city; for that the symbol of the owl would have sufficed. Her image is also a reference to the world of gods shared by the entire Greek cultural sphere. With respect to this part of the coin's form, observers from outside of Athens were thus included within the sphere of communication.

Roughly around 480 B.C., a small lunar crescent was added to the owl and a diadem of olive to Athena's head, "thus establishing designs which were to remain unchanged (apart from stylistic developments) throughout the fifth, fourth, and much of the third century."[34] Part of these "stylistic developments" are an increasing simplification of Athena's distinguishing features and the slow expansion of the reverse's incuse square beyond the coin's edge. The corners of the mark disappeared. The mark had become something taken for granted.

Hitherto the success of the owl coinage and its remarkable stability were attributed mainly to factors unrelated to payments. Athens had silver resources through tributary payments and the exploitation of the silver mines at Laurium. Moreover, documents such as a decree from 420 B.C. indicate that the city sought to assert her currency with power and violence vis-à-vis allies and those owing tribute. The arguments, however, remain unsatisfactory. Attempts to monopolize the money supply, such as the one mentioned above, were frequent, yet in each instance they were sustainable only for a short time. What is more significant is that Athens's demand for currency was seemingly independent of available silver reserves. Athens's "hunger for money" grew to the point that the treasure of the Delphic League was taken to Athens and struck into coins. It comprised 5,000 talents equaling 125 tons of silver or 7.5 million tetradrachms. Athens also regularly took in silver revenues of a similar amount. With the money supply thus generated, the construction project of the Acropolis could be financed. Thus it was not the trade volume of commodities that required the creation of a suitable means of payment. Rather, a means of payment could be stimulated by identifying a suitable external referent, for example, the construction project. As a result, this money, once brought into circulation, was then also available for the payment of common goods. A good deal of the minted coins left the circulation of the Athenian economy and became the means of exchange for the nondomestic economies from Asia Minor to Afghanistan. The popularity of the owl coinage was additionally supported by local imitations and by adaptations especially in the Persian sphere of influence.

Hippias's reform had effects that he as an observer could not have predicted. There is reason to believe that Athens's rise as the financial and trade center of the Greek world is at least partly due to the dynamics generated by the reproduction requirements of the new coin form. These included the emergence of financial institutions that were novel enterprises specializing in exchanging and storing monetary forms. For the first time, a complex of institutions took shape within society's communication of payment which dealt solely with the reproduction of the monetary medium. The monetary economy, hitherto limited to special circles and a few cities, spread quickly;[35] minting sites emerged all over Greece.[36] A new epoch in the evolution of European history began.

Unresolved Issues

With our observations of Athenian coinage, we have entered our own cultural space and thus more familiar territory. In the previous section it was unnecessary to explain what the "construction project of the Acropolis" referred to. We also have reached the limit of questions that can be answered with existing source materials. Beyond this boundary, that is, outside of the coin form, questions arise that in the context of our investigation can only be illuminated tentatively.

What is especially striking is a peculiar oscillation of the signs' referentiality in the evolution of coins. On the one hand, such signs refer to private use that becomes credible externally, that is, in communication with other closed systems. On the other, the signs refer to public use that is legitimized within the closed monetary system.[37] The evolution of the coin began with private signs (form 1 and form 2). These signs enabled payments outside of a credit community. The payments were used in turn for a few expensive consumer and investment goods, possibly also for services purchased from external sources.

The credibility of the marked nuggets arose mainly from their recognizability. The sphere of circulation around the minting household, though very small at first, grew rapidly when, with form 2, the previously discussed double ambiguity emerged. With form 3 the information content of the coin essentially shifted to the obverse, while the reverse with its marks preserved the continuity of the established use. At this time the obverse exhibited a public image—the totem animal. Here again an ambiguity proved of service: the religious and political meaning gained through the image was interpreted economically and thus excluded risks that could not previously be ruled out. But internal payments with the new coin form by no means penetrated the entire ongoing economy. The coins were employed for new kinds of payments that arose out of communication with central ruling households, namely taxes, tributes, tariffs, and reimbursements for corresponding expenses of the ruler. Again, the payments occurred at the periphery, this time, however, on the inside of society. The dissemination of the new coin form in Ionian cities enabled in turn an increase of external trade. Although the nominal value of the coins began to diverge from their metallic value, the trust in their acceptance as a means of exchange had grown sufficiently to sustain their use as payment. In form 4, the Aeginetan currency, nothing but the chemical, metallic form of the coin changed. Still, it was this currency that succeeded in expanding the internal use of money, for the use of silver enabled the introduction of smaller denominations. That in turn led to the extension of the external use of the coinage in open trade of mass commodities.[38] Again, the new forms of payment occurred on the inside of communication systems that employed them. With form 5, however, the signs migrated once more. The symbol that was hitherto public now denoted a minting site, that is, a private organization external to the bearer. The public totem symbol was replaced by a deity interpreted as a human head. Thus religious and aesthetic artistry was employed anew to secure economic credibility. The new coinage was carried to the periphery of the Greek culture. It did not necessarily return to those who initially coined it, but circulated freely within an economic area that had become immensely wide and dense.

The coin form thus founded remained stable until modern times. The reasons for this stability are not clear. One hypothesis proposes that the nesting of forms prevented their dissolution: First, the use of precious metal delimits one (chemical) form from all others; second, images set the coin apart from all other material forms; and third, images of humanized deities distinguish cultural self-observation in the form of widely known gods from such self-observation rendered through local totemic forms. Another hypothesis suggests that form 5 succeeded in accommodating the oscillation between private and public markings in one form. The reference to the mint created credibility in that it connected the coin to an origin, that is, a private starting point of positing value; the reference to the entire culture created credibility by making the communicated value collectively understandable. Both factors—another ambiguity—were now united in one semiotic form. In the interplay between public mints and private banks the coinage was established and disseminated. Only in this context did the value standard for different monetary forms (objects, accounting entries, and credit) become uniform.[39] This, of course, applied only to the Greek economy. In the surrounding societies, especially in the Persian empire, the use of coins remained limited to the trade in luxury goods; in the Roman communication of payment it became established only in the third century B.C. Moreover, not until the eighteenth century, when paper bills emerged, did a form of monetary exchange take root that was largely independent of political legitimization and material valuation.

Not only does the oscillation of the marks remain for the most part an unsolved matter, but the reasons for the periodic emergence of human heads with different meanings are also unclear. More than a hundred years prior to the Athena images there were human heads and busts on early Ionian coins. Presumably these images, like those of animals, are to be interpreted as "spirits" with local religious significance.[40] Thus the signified entities were essentially nonhuman, and their human form was rather a matter of chance. In the case of Athena and the other Olympic gods, however, matters changed: now the world of the gods was a meticulous representation of the human world. It is this cultural achievement that enabled the Greeks to articulate a closed yet highly differentiated understanding of the world. The ambiguity in the depiction of a head that is recognizably human yet interpreted as divine is precisely the paradox that closes off the cultural sphere observed in the Greek way. Two hundred years after the first owl coins, heads appeared that were perceived as human. The tradition started with the head of Alexander the Great. Very gradually the image of the ruler was uncoupled from the familiar representation of the divine Heracles.[41] Thus the beginnings were once more marked by an ambiguity, and the transition was as form-preserving as possible. That images of rulers were imprinted so late might have stemmed from the inhibition to do so. It is more likely that Alexander was the first ruler sufficiently popular and widely enough admired not only to compete with a Greek god but even to be venerated above the gods in some areas of his empire located outside the Greek cultural sphere. This then made the establishment of the new monetary form possible.[42]

We now leave the observations that can be obtained from a historical perusal of the evolution of coinage. Before exiting the text altogether, however, we shall consider the methodological consequences for observing the contemporary economy and its monetary medium. Today we are just as bound to the forms of the utilized media of payments as was the case at the time of the Lydian merchant households. We trust bank notes, credit cards, and special drawing rights, just as the traders trusted the marked *phthoides*. However, we cannot observe the forms of the structures of expectation since we ourselves as addressees are part of the network. With respect to coins we had the advantage of observing communication of payments via material forms. In the contemporary communication of payment, material forms have largely become irrelevant:[43] the creation of money has become more and more the pure creation of credit by a complex organization, that is, the network of financial institutions.[44] Instead, we have today a different source for observation available: We can register and reconstruct the paths that communication of payments takes. "Being a financial institution" is not necessarily bound to a particular type of organization. Insurance companies, pension funds, and all enterprises big enough to format flows of payments by themselves are part of the communication between financial institutions. The monetary medium is reproduced today by a multitude of organizations. The oscillation between private sites of money creation and trust in a public (societal) monetary community[45] symbolized by central banks still occurs. But it does so in a very complex structure, a fact we can only briefly mention here.

This perspective allows us to pose new questions about what can be taken for granted in today's economy and its monetary medium. We have only to interpret observed semiotic events as autonomous observations that of necessity relate to both previous and future observations. This applies equally to both economic and scientific observations of the reproduction of observations. Subject and object have disappeared; they are dissolved in the distinction between the communication game called society on the one hand and its environment on the other.

RUDOLF STICHWEH | **THE FORM OF THE UNIVERSITY**

I

Does *the* university really exist? Given the extreme diversity of institutions of higher education in contemporary societies, does it still make sense to call some or all of these organizations "universities"? What underlying judgments and partisanships are involved in any such attempt? The 1973 theoretical study of the university by Talcott Parsons and Gerald M. Platt[1] suffered criticism and neglect in part because the authors entitled it *The American University* even though it focuses only on a small group of schools of a specific type: predominantly private undergraduate colleges, professional schools, and graduate schools with a definite emphasis on research.

Two contrasting points are to be presented here: On the one hand, we can observe the astounding continuity of the name "university" and the remarkable persistence of individual universities.[2] On the other hand, the contemporary situation prompts us to ask whether a common intellectual-organizational core exists in the multitude of higher-educational institutions, by means of which this plurality can be comprehended.

I opened with the question: Does the university really exist? To help identify and describe the university, I will recast this question in a twofold way: How does *an external observer* distinguish the university from other organizations? And how does the university distinguish *itself*? This reframing has the advantage of explicitly recognizing that the elicited distinctions involve processes of (external) observation and self-observation in a social environment. Furthermore, reframing the opening question—which aimed to determine the identity of the university directly—as questions regarding a *distinction* and *its criterion* permits us to differentiate the university from something else,[3] whether from any other phenomenon (university vs. world) or from more narrowly specified phenomena (university vs. school). Any distinction has two sides; these can be varied independently of each other when only one side of the distinction is replaced at a given moment, for instance, when the university is distinguished from the *Fachhochschule*[4] but no longer from the high school. At the same time, however, the meaning of the other side of the distinction is thereby inevitably altered as well; hence, the two sides are not truly independent of each other.

We can therefore expect that our considerations will typically have to deal with two intertwined distinctions. First, the university is distinguished from something else that is not a university, and second, this distinction between the university and something else is drawn by means of yet another distinction. For example, in the Middle Ages one could make the distinction between university and monastery by distinguishing between the practice of religious faith, obliged to devotion, and theological discourse, to which a comparable commitment to devotion pertains only *after* the scholarly problem is solved.[5] Changes in both the *object of the distinction* (university vs. monastery, vs. school, vs. profit-oriented enterprise) and the *criterion of the distinction* (devout/scholarly, utilitarian/scholarly, scientific / "scholarly only") are then of historical as well as sociological interest.

Determinations that an observer makes by choosing distinctions with two sides, opting for one of the sides, and possibly varying one of them (and perhaps subsequently observing the unity of the employed distinction by means of other distinctions)[6] are today often called *forms*.[7] According to this terminology, the following considerations address the form or (probably more appropriate for the present situation) *forms* of the university. The issue always concerns distinctions that allow *operations* to be connected yet that are already operations themselves, since temporarily opting for one side of the distinction and making the distinction are performed in one and the same act. Subsequently, many connecting operations are possible: affirming the momentarily chosen designation for one side of the distinction; replacing it with its conceptual opposite; internally differentiating the one designated side by way of another distinction; moving to the other side of the distinction; seeking a comprehensive unity of the distinction that ultimately turns out to be yet another distinction.

II

For the European society of the late Middle Ages and early modernity the question of the university's constitutive difference from other institutions must have been relatively unproblematic. For centuries *universalism versus particularism of knowledge* evidently functioned as the guiding distinction,[8] implying the university's systematic preference for knowledge not narrowed by any restrictions on its use and not attached to any index to region or social rank or other particular reference. At the time the European university emerged, this preference for universalism probably originally implied a delimitation from schools of religious orders ("particular studies") and theological seminaries of only regional importance. Independently of this historical origin, the distinction between universalism and particularism turns out to be a stable self-identification of the university at least until the end of the eighteenth century. Given this distinction, it was logically consistent, for instance, that the English universities of early modernity did not teach a legal system of only particular relevance, namely "Common Law," but held fast to the universal "Civil Law." Similarly, the German academic law faculties even of the nineteenth century ignored or marginalized territorial legal codes in their academic doctrines, holding instead either to the Roman *ius commune* tradition as systematized in the seventeenth and eighteenth centuries or to the "science of Pandects" as a "system of today's Roman law."[9]

The distinction between a universalistic and a particularistic handling of knowledge, however, can hardly serve to determine whether a particular organization is a university or a school of another kind. In late-medieval and early-modern Europe a twofold operationalization of this guiding distinction took place. First, through the difference *privileged/nonprivileged*. Universities rested their claim to universalism on the two other universal medieval powers—the Catholic Church and the Roman-German empire—which assigned them this universalistic mission and furnished the appropriate entitlements (e.g., immunities and privileges). Thus universities gained distinction through the difference *privileged versus nonprivileged*. Second, one of the decisive entitlements granted was the power to award academic degrees. Moreover, those degrees were claimed to be valid all over Europe. Thus the difference between universities and other schools lay in the fact that the former were allowed to confer degrees. Any potential university student could certainly opt for one of these other schools (knightly academies, "academic schools" [*akademische Gymnasien*],[10] seminaries), which might better impart practical knowledge, cosmopolitanism, and competence relevant to social rank, but he would thereby forgo the degree and its universal validity.

Since our issue is the modern university, we omit—though it is, in principle, required—detailing and differentiating these accounts historically. It was only important to indicate the relative simplicity and the historical stability of a model that emerged in the late Middle Ages. Concerning stability, one must concede that the early-modern university of the sixteenth to eighteenth centuries certainly came increasingly under the influence of utilitarian expectations, interpreted in a particularistic manner; privileges became unobtainable or were no longer sought (owing to denominational schism and territorial sovereignty); and degrees were granted by institutions that lacked the proper privileged status. Nonetheless, it is crucial to acknowledge the persistence of the model in two respects.

First, for the aforementioned examples of instability one can always find counterexamples, such as the previously mentioned resistance of law faculties to territorially specified legal systems. Later cases in which universities won privileges also serve as counterexamples—for instance, the Protestant university of Göttingen, founded in 1734, sought and was granted privileged status from the imperial Catholic Court Council (*Reichshofrat*) in Vienna. A final counterexample is the case of Dutch universities that granted "unauthorized" degrees but strove to compensate by adopting elements of the graduation ceremony and attempting to surpass the privileged universities in scientific achievements.

A second observation regarding the continuity of the guiding distinction between universalism and particularism is of equal importance. Despite the weakening, provincializing, and pragmatizing of the old model, a new one for distinguishing the university did not emerge until the end of the eighteenth century. Therefore, either one did not distinguish at all or one had to continue using the old distinction. The universalism/particularism distinction thus also survived owing to the absence of an alternative.

In early-modern Europe, the distinction between universalism and particularism confronted a major challenge: the invention of the printing press. As an inevitable consequence of this invention, the universality of knowledge can be properly represented only in the multitude of currently available books and can be appropriated and continued qua universality only in that manner. This situation changes the status of academic teaching dramatically. The university instructor is no longer one of the few who knows and owns the small number of available books and who reads from them—at best explicating them—to his audience. Instead it becomes apparent that academic teaching is a particularistic form of access to universalistic knowledge since it involves a *person*. This brings about the very concept of *study* for the first time, a concept that in principle makes the instructor's oral communication, with its blend of particularistic and universalistic elements, distinguishable from learned communication addressed to other scholars and transformed into writing.[11] A passage in Richard Mulcaster succinctly expresses this discontinuity but also the disregard of its consequences. He stresses the readings, by that time considered essential, of the university instructor (still called a "reader" or "lecturer" in British English today), and analyzes the student's situation: the student has "in one lecture the benefit of *his readers universall studie*, and that so fitted to his hand, as he may *straight way use it, without further thinking on.*" A few lines down Mulcaster summarizes the achievement of the university instructor and its inherent economy for the student: "his reader is his library."[12] Mulcaster emphasizes the universality of the instructor's studies while minimizing the instructor's impact on the knowledge he imparts in order to maintain that this knowledge can be utilized by everyone without any intervening reflection. Here I can only note that the radical challenge to the guiding distinction between universalism and particularism by the printing press would make a fascinating research topic. I suspect that the early-modern university covered up this problem in various ways, and that it was only the academic revolution of the early nineteenth century that deliberately confronted this problem in, among other things, a theory of the academic lecture.[13] However, this is a remarkably uncharted territory, so I only want to mention it here as a problem.[14]

III

It was certainly plausible to view the development sketched above as one that would ultimately result in the *abolition of the university*, and many observers and participants during the years around 1800 did make precisely this diagnosis. What could be observed was a gradual erosion of the former constitutive distinction between universalism and particularism of knowledge and a refusal to reflect on the radical challenge the printing press posed to this guiding distinction. Both problems seemed to point to the death of the university. Furthermore, after the establishment of the nation-state and as a result of the attendant definitive nationalizing of the higher-educational system (even in the newly founded American Republic the first six presidents discussed the creation of a "National University" in Washington D.C.),[15] it became evident that a direct continuation of the universalistic old-European model was no longer conceivable. But what does one do with the university once one realizes that it has already become a "self-substitutive order"[16] possessing an indispensable function?[17] It can no longer be abolished, nor can newly founded universities readily take the place of the old ones.

What is left, then, is to reform the university. From the beginning of the nineteenth century up to the current discussion, the topos *reformed/unreformed*, or alternatively *reformable/unreformable*, has been decisive in external observation of the university and in the university's self-observation.[18] And the included excluded third of this distinction is without any doubt *the state*. One can be of the opinion that universities, like other corporations, are by themselves incapable of internal reform and thus have to be reformed from without.[19] Alternatively, the university itself may signal its wishes to reform to the state, in order to overcome internal opposition or nonnegotiable statutes by means of an external authority, a recurrent situation at Oxford and Cambridge universities since the mid–nineteenth century.[20] The university can certainly also carry out unrelenting reform from within in order to preempt state intervention. Finally, a new institution might conceivably be introduced in lieu of reform of the university on grounds that the university is incapable of reform. As evidence we cite here the former president of the West German Science Council, Dieter Simon, who publicly declared that in the last twenty years universities have proven to be "incapable of reform" while *Fach-*

hochschulen have implemented "the most successful reform of academic studies."[21] What is striking in this statement is that the university/*Fachhochschule* distinction (excluding the *Fachhochschule* from the higher-educational system) is replaced by the distinction (meant to be internal to the academic system) between those incapable of reform and the reformed, and that in shifting distinctions (that is, by including a new institutional type of higher education) the boundary of the university is redrawn.

This seems to be a general implication of the reformed/unreformable distinction. Hitherto dominant university models can at any time be rejected as being not yet reformed or unreformable. They are contrasted with a reformed university, which in effect amounts to a broadening of the concept of the university. Furthermore, we should not forget that within the spectrum of reformed/unreformed all options or designations have the purpose of *either drawing the state into, or keeping it out of, the university*. The German neologism *Reformuniversität* (reform university) indicates this quite well: the term presents an entire organization as an incorporated reform while expressing a sort of self-authorization and self-commitment of the state both to intervene in and (by suspending some of its own rules) to stay out of the university's affairs.

What objective is pursued in reforming the university? And how can we make certain that the university, once reformed, is still a university and not merely a subordinate state institution? A classic answer to these questions exists, but it apparently was not imaginable before the beginning of the nineteenth century. The discourse on reforming the university is evidently older. Between the sixteenth and eighteenth centuries, *reformations* of the university (mostly by the territorial ruler) occurred, in many cases presupposing the early-modern concept of reform as an incremental remedying of problems that had crept in. In this respect, what is new and specifically modern regarding the distinction we have just analyzed is only the intensification of the duality reformed/unreformed, excluding any other options. By contrast, only after the turn of the eighteenth century was it possible to declare that the *reform objective is the realization of the idea of the university*. Instead of fixing problems, the reform thus seems now to be about value implementation in a Parsonian sense. For this an increasing distance between value judgments and descriptive statements about reality figures as one precondition. There are two loci classici for the idea of an "idea of the university": the Berlin lectures (1808/9) by the Dane Henrik Steffens, a romantic convert, and the Dublin lectures (1852) of John Henry Newman, an Englishman who converted to Catholicism.[22] Both texts have received so many revivals that Michael Cohen and James March in our day could conclude ironically: "Almost any educated person can deliver a lecture entitled 'The Goals of the University.' Almost no one will listen to the lecture voluntarily."[23]

An idea is not a distinction. Yet it presupposes a distinction, namely that between *idea* and *reality*.[24] The reality of the university, however, is not the reality of its idea; it is always somewhere else. The internal dynamics of the university's evolution in modernity transformed the conception of an "idea of a university," initially signaling a constructive intent, into a defensive conception. Now the issue is admonishing values deemed forgotten rather than genuine implementation of values. The wish to reinstate the late-medieval distinction between universalistic and particularistic is noticeable as early as Newman: "The view taken of a University in these Discourses is the following: That it is the place of *teaching* universal *knowledge*."[25] Insofar as the modern university probably never had a premonition of what would become of it, as the following considerations will show, its evolution cannot be reasonably viewed as the result of implementing an a priori idea, that is, of fully utilizing the self-generated cultural substance of the university.

Newman's idea of transmitting universal knowledge became untenable because it presupposed assured knowledge at a time when the certainty and determinability of knowledge was becoming questionable for many. This new experience of questionable knowledge, according to Owen Chadwick, goes back to the British university of the 1850s: "To receive assured knowledge, that had been the ideal. But one of the new elements in assured knowledge was the proposition that less knowledge was assured than had hitherto been supposed."[26] In the evolution of the German university, the same insight and experience was expressed after 1800 by means of a distinction that proved to be more influential than its encouching discourse of an "idea of the university": Academic teaching had to be *scientific*, which is precisely what makes the university distinct from schools and other educational institutions.[27] In effect, three or even four distinctions are linked here. First, in an abstraction transcending school types, which was not yet possible in the eighteenth century, *the school* is more clearly than ever before the frame by which the university delimits itself. At the same time the university is assigned to the educational system, and the school/university distinction is analyzed as an internal distinction of the educational system.[28] Second, the difference between school and university cannot be marked by means of the universalism/particularism distinction as it could in the late Middle Ages. Instead, the universalism of the university—as opposed to the also assumed universalism of the school—is now qualified by the imperative to subject universal knowledge to scientific treatment.[29] What this imperative means precisely can only be explicated with the help of yet a third distinction. Knowledge is to be handled by a mind that is always ready to suspend anew the question of the *truth or untruth* of knowledge and that considers attributions to either side as provisional and revisable in light of new evidence. Insofar as this can lead to reflecting on the limits of cognition, the new definition of the university as a scientific university reaches only tentative closure in a fourth distinction we might call *cognition/reflection*. In 1925, Carl Heinrich Becker regarded this reflexive turn as the decisive feature of the German university's evolution since Kant. It turned "the relentless enthusiasm for cognition daunted by nothing into a critical stance toward the process and possibilities of cognition."[30]

What becomes apparent here is that the emergence of a form within the university, that is, the co-utilization of the form of science (the binary code of truth/untruth) by academic organizations that are at the same time marked more clearly as organizations of the educational system, has been proven to be an irreversible, radical shift. I do not want to elaborate on this relatively well known process; let me merely point to another, far-reaching consequence. The scientific nature of university education does not equal a socialization of scientific *results*. Academic education would then differ from the popularization of science only in scope and systematicity, but it would not engage in science as a process of producing knowledge. In that respect the postulate of the *unity of teaching and research*, meaning above all that the *difference* between teaching and research may not be used in the university as a form-giving device,[31] is a strong implication of the scientific university.[32] It would therefore be interesting to compare the creativity of contemporary university systems in achieving the unity of teaching and research within teaching settings where this is deemed particularly unlikely to succeed.[33]

IV

A distinction that has been of equal importance to the evolutionary dynamics of the modern university (though not to its self-description) as the issue of the university's scientific nature, and latently or manifestly an issue in all university reforms, is the matter of *including versus excluding* prospective students. From the eighteenth through the twentieth centuries, the European university adopted many inclusionary policies with some success: rank- and class-based preconditions for entering universities were reduced; admissions requirements were differentiated to adjust for dissimilar high-school educations, a measure that followed the initial, early-nineteenth-century problem of homogenizing the beginning of school education;[34] women were admitted to academic studies; grants were issued to students from low-income families; and universities were regionalized, a development that enhanced the access to university education of students with local or regional ties.[35] As a result of these inclusions, every man and woman now can be viewed as a potential university student. Yet the European universities have never abandoned the idea that from the pool of high-school graduates, only a few should be admitted; less ideally, many might be accepted, but never should all of them be granted admission. Clearly the social structure presupposed by this idea reproduces a pattern of social rank and generates (and regenerates in every generation) an "estate" of the learned, educated, and professionals within the social system, but this has never affected the described structural preference. Many self-descriptions are about the formation of social elites, and an elite is, as Joseph Ben-David pointed out, for a European intuitively small in number, whereas for someone in the United States it is more numerous. Even if one were to concede that someday 20 percent of a high-school graduating class could be a university admissions figure appropriate for the existing pool of talent, one would also believe that one had to draw the line at 30 percent and that it would be possible to maintain such a position over time.

I hypothesize that the European universities have never really made *inclusion* their agenda. It is true that they took inclusion into account to some extent by reducing many extraneous barriers to attending the university, yet they have never programmatically embraced the idea of offering everyone interested a college education and in addition have not proactively encouraged interest. Instead, though conceding in principle that talent should be fully utilized, they also regard the concept of talent as a viable means for limiting access to higher education. Consequently *admitting / not admitting* becomes the tacit leading distinction of universities. The unity of this distinction presumably is *admitting*; that is to say, admission restrictions are always only feasible for a limited period of time, after which politics and universities are faced with developments they never really intended. One is always disappointed in one way: it is always too many rather than too few students that enter the university.

It is interesting to examine the same quantitative concern on an institutional level. First, one typically encounters the assumption that there ought to be only a small number of universities. (The ideas about what a small number is can vary, but the structural preference itself does not.)[36] The multitudinousness of North Rhine-Westphalian *Gesamthochschulen*[37] is an acknowledged political blunder. A similar instance occurred in the revamping of the university system in the former GDR. Apart from local political and economic interest groups, those involved in the revamping effort generally agreed that there should be only a few universities—and this despite the known fact that a steady growth of metropolitan universities entails unbearable infrastructural pressures.[38]

A second aspect of concern with quantity at the institutional level is exemplified in France and Germany in particular. These nations have time and again attempted in the nineteenth and twentieth centuries to channel students into newly created subacademic institutions.[39] Decades later one then has to grant these institutions transfer entitlements for their students, and after that the right to award Ph.D.s, and finally university status; one thus has to include them in the university system. Without doubt, innovative institutions of higher education can emerge in this way. Yet it is noteworthy that these new institutions result not from a concern about new forms of knowledge and a corresponding intent to give shape to them but rather from a motive of exclusion. Something comparable can be observed in the continuous German discussion on the expansion of *Fachhochschulen*.

Another way in which the university might operationally handle the issue of inclusion and exclusion is simply to admit all students, without, however, making any concessions on curriculum and course requirements. Friedrich Paulsen described this scenario using the distinction between a *freshman mentality* and *science*, justifying the university's practical disregard for a considerable percentage of the existing student body: "Overcoming the mere freshman attitude vis-à-vis science is still the common demand, though we know that not all students are capable of meeting this standard. Although in some ways we could be more accommodating, we are not so because we do not consider this beneficial if it has to be done at the expense of those more advanced."[40] Contemporary systems of higher education that want to deal with the universality of access in a different way have to replace the normative expectation of their scientific nature with the "value-added" principle of evaluation,[41] meaning that there is no longer a normative standard for defining a successfully completed academic education. The one still universally applicable principle of evaluation is thus whether or not the students have gained something from their studies.

V

In the respects just indicated, only one of the major higher-educational systems in the world, namely that of the United States, followed a different course. From the outset the U.S. has had numerous colleges and universities, far exceeding the number in each traditional European system of higher education. In 1860, at the beginning of the Civil War, there were already 250 colleges, of which 180 still exist.[42] By 1910 the number had risen to almost 1,000 colleges and universities, with a total student population of one-third of a million; at the same time, France had 16 universities, with a total of 40,000 students. The obvious objection that the French universities are to be regarded as institutions of incomparably higher academic standards is still justified for 1910, yet it is weakened by the fact that in 1910 the faculty of the American higher-educational institutions alone came very close to the number of French students.[43]

In this regard, the American system of higher education is a classic case of *polycontexturality*.[44] All distinctions used in other countries for marking the university's boundary are instructive for analyzing the internal differentiations of the American higher-educational system.[45] Let us look at a few examples. (1) *Scientific/nonscientific*: In the U.S. the demand that university education be scientific did not lead to a transformation of the preexisting university system but rather was implemented by means of new institutions (e.g., "School of Arts and Sciences," "Graduate School").[46] (2) *Secular/religious*: Since the internal structure of the American higher-educational system from the outset reproduced the plurality of imported denominations and religious beliefs,[47] the division secular/religious was never able to function as the boundary of the university system, whereas in France, for instance, this division established a boundary in the educational system that for a long time was marked much more clearly than the distinction between the secondary school system and the higher-educational system. (3) *Private/public*: In this respect as well, the American university system represents social plurality, for in the U.S. no one is prohibited from founding a higher-educational institution, and hence the state and entrepreneurs have competed with one another. (4) *Unitary location / federal structure*: In the American (that is, campus-style) university, on the one hand, unitary location, allowing spatial and architectural connection among the university's component parts, is a familiar feature by which the union of a large number of disciplines and programs is expressed.[48] On the other hand, the American university has produced, through state universities, federal structures combining many separate campuses into one organization (e.g., the University of California System).[49] (5) *Founding/preserving* of organizations: Perusing the history of the American higher-educational system, on the one hand, we can observe a preference for founding new universities, that is, for replacing organizations that have become inefficient, analogous to the way an economic system operates.[50] On the other hand, we find an imperative of preservation at work for institutions that have gained historical significance, meaning that the motive for upholding a tradition is apparently institutionalized as well.[51] (6) *Liberal arts / professional education*: Especially on the level of two-year colleges, this distinction is striking. While some endeavor to provide a liberal-arts education possibly leading to studies at a research university, others are primarily concerned with a professional education that shortens the time spent in the academic sector.[52]

A good indication of the polycontexturality of the American higher-educational system is the fact that school and university types have remained semantically fairly undifferentiated: "school," "college," "university" (unlike the terms in British English) can still be used interchangeably today.[53] It is furthermore noteworthy that these terms exhibit no correlation to the distinction between private and public institutions that is otherwise central to the American higher-educational system.[54]

The multiple distinctions that operate in the American higher-educational system are employed to observe differences not only *between* the various types of academic organizations but also *within* one and the same institution. Martin Trow, to whom we owe the best analyses of the *simultaneity* of elite education, mass education, and general education in the American university system, repeatedly emphasized in his studies the observation that all three types of education coexist in large state universities.[55] This fact, by the way, illustrates a pattern of structure formation that can be observed in social systems at large. New structures do not displace their preceding structures; rather they join them and complexify the types of structure formation in a system. A key question for any organizational analysis then becomes how an organization can successfully consider a plurality of legitimate distinctions both independently of and equally with each other.

VI

The central analytical terms in recent discussions of the university system's capacity for considering multiple distinctions are "loose coupling" and "decoupling."[56] Common to all analyses of this kind is the idea that real systems have a "dialectical"[57] and "paradoxical" make-up. That is to say, apparent contradictions may occur in systems simultaneously or sequentially, contradictions that are enabled by the system's process of differentiating (most importantly, internal differentiation). Another commonality of these analyses is their focus both on *couplings* (interactions) between a system and its environment and on couplings (interactions) internal to a system, and their attempt to find a terminology that would encompass and illuminate connections between these perspectives. The autonomy of a system would then consist in balances between those two dimensions. For instance, a high degree of *responsiveness* of a system to its social environment is compatible with the system's autonomy only if the responsiveness is offset by a large number of *system-internal distinctions*. This balance allows the system to locally limit every specific response to concrete and possibly new environmental expectations. Because of this localization of effects, the autonomy of the system is not affected. Any new environmental expectation, as soon as it is transformed into a system response, takes on the form of a new, institutionalized difference, which, since it is joining many preexisting relevant distinctions, only partially restructures the system and does not affect the system's internal autonomy in connecting new distinctions to hitherto effective distinctions. *Responsiveness* and *polycontexturality* of a system are therefore to be understood as coevolving and codependent principles.

We can identify this construal of a connection between externally oriented responsiveness and internally open polycontexturality as the common intellectual backdrop of the various theories of loose coupling promoted today. It also informs John W. Meyer's and Brian Rowan's suggested distinction between *formal structure*, that is, the official, ceremonial, externally oriented side of the university, and *activities*, meaning the real events within the university, the actual day-to-day work activities that can hardly be controlled by the low-informational formal structure and that exhibit a much higher degree of complexity.[58]

The relationship between formal structure and activities is surely not to be regarded as one of complete dissociation, as if the formal structures concerned only myths and ceremonies designed to lend some semblance of formal rationality to nonrational events in universities.[59] This would imply a decoupling of the university from its environment and a complete setting free of internal university events, since the ceremonial self-presentation is devised for the environment and varies independently of internal events, absorbing all environmental pressures. Such an assumption would fail to recognize, among other things, that the formal structure of the university is directed not only toward the outside but also toward the inside. After all, there is likewise a demand within the workings of a system for unifying formulas to ensure that an increasingly polycontextural system is in fact still *one* system. Given this situation, it makes sense, I believe, to interpret the relationships between formal structure and activities in light of the theoretical distinction between *medium* and *form*. This theoretical paradigm should also enable us to answer the question of how a university determined by a multiplicity of differences can still be perceived as one system, and thus how the unity of the system is possible.

The distinction between medium and form refers to the fact that stronger form configurations with invariable links between some of the system's elements emerge out of a very large set of its elements which are only loosely connected to one another.[60] What serves as the condition for the potential emergence of such forms we then call medium. This makes the concept of medium somewhat equivocal. For one, it can refer to a plurality of loosely connected elements that are viewed as a formable medium. But it can also mean a stock of symbols independent of the system's elements. This reservoir of symbols can be correlated with basic operations of the system to enable the construction of complex structures as links between systemic operations/activities and symbols that become operations of the system themselves.

In this sense, is there a *medium of higher education* that allows for the connection of formal structure and activities and succeeds in this effort because a stock of symbols enables the emergence of form via basic teaching and learning activities? Another theoretical expectation we need to consider in answering this question is that we ought to be able to interpret a possible connection between medium and form from the form side as well. That is to say, we also need to ask how the introduction of new forms enables one to regulate and vary the basic teaching and learning level. The question of whether or not a medium of the higher-educational system exists has often been answered in the negative since, so it has been pointed out, educational activities depend on interaction with other systems. It is argued that while the predominance of interactional systems allows action to be professionalized, that is, allows sophisticated techniques to be developed to deal with the uncertainty of future situations of interaction, nevertheless that predominance does not permit the formation of a medium connecting decisions in one system of interaction with those in another. One objection to this standard argument comes immediately to mind: there seems to be an obvious candidate for the medium of higher education. Since the rise of the European universities, academic teaching and learning have been geared toward *examinations*, which on the one hand indicate the completion of a program of study, a study phase, or an academic year, and on the other hand constitute an essential element of the external representation of the university. One might just think of public disputation as an example where the two aspects are united in one act.[61] Additionally, examinations serve to permanently validate the processes and results (credits, certificates, degrees) of university education in the world outside the university; degrees are hence a form of effectively stabilizing the results of university education externally.

In this sense, degrees, examinations, and course certificates thus were initially a stock of symbols intended for special situations. A few educational systems, for instance the Jesuit seminaries spread over Catholic Europe, tended to universalize the connection between educational activities and the symbolic reinforcement of comparing achievements—that is, they tended to institutionalize competition and combine every school activity with assessments of achievement. The American colleges and universities of the nineteenth and twentieth centuries began to establish a permanent link among three components: subject areas of knowledge, courses, and evaluations (grades). The linkage goes as follows: (1) The curriculum must comprise subject areas; (2) these subject areas must be decomposable in such a way that they can be institutionalized in the form of courses; and (3) these courses must warrant an evaluation of participation formalizable as grades. At the onset of the twentieth century, Thorstein Veblen could still view this tripartite configuration (subject matter, course, evaluation of performance) as an expression of the penal quality of the American university, in contrast to the academic freedom of European universities. Veblen described this special American situation as a predictable educational counterpart to a culture that for him tended to conceive young adults' sincere passion for knowledge as a form of social deviance.[62] His critical assessment has most likely become obsolete owing to parallel developments in the European university systems since then.

I therefore would like to suggest analyzing the academic *credit* system (*Scheine* in German, i.e., certificate for the completion of course work) as the medium of university education.[63] Without delving into a detailed discussion here, I merely want to note that the function of credits—as a suitable medium allowing the emergence of increasingly complex forms in university processes—can be more closely defined in several ways:

1. The focus on the rewarding of credits regulates the selection of academic subjects and the subsequent decomposition of the chosen subjects into subtopics for which credits can be given.

2. For those enrolled in a course, credits regulate the motivation to participate; they direct interest in certain subject matters in specific directions, and they generate a kind of conformity to the intellectual premises of the university. The latter is true because for a student to receive credits he or she, without being exhorted to share the opinions of the instructor,

must at least not deem the academic course topics and the method of processing questions completely irrelevant (or he or she at least must not communicate this opinion).

3. The fact that credits have motivational effects on both sides ties together teaching and learning into one system. The university instructor who is willing to specify his or her course topics and break them down into units for credit receives in return participatory motivation on the part of the students, their willingness to sign up for activities, and, as a result, a sort of assured minimal success of the instructor's courses.

4. In the credit medium, then, an arrangement of complex forms is possible that may be described on the teaching side as *curricula* and on the student side as *careers of study* marked by classes either taken for credit or avoided.

5. Curricula and standardized university careers together define a kind of *formal structure* of the university. They have in common that they are descriptions behind which actual activities can be hidden that one would not have readily perceived just knowing these descriptions. Conversely, it is also true that these formal-structural components are not compatible with absolutely any activity whatsoever. Rather, one has to assume that they impact on activities in a restrictive or selective fashion, though they are not able to *instruct* them in a precise sense.[64] This condition provides the chance to influence the structure of university activities across several levels by introducing new forms (curricular distinctions, structures of faculties, subjects, and disciplines). Here we have to note that the distinction between medium and form is relative; it pertains only to the indicated level of a system. What in view of one chosen systemic level is a form may possess on the next systemic level up properties of a medium; thus it appears as an element that needs to be structured.

6. Credits permit transcending boundaries in a polycontextural university system that also consists of many organizations. This is possible because one can translate the unmanageable matter of comparing incommensurable study programs at different higher-educational systems into the practically manageable issue of accepting / not-accepting credits.

VII

Academic credit as a medium of university *education* is only one of the modern university's functional relationships. The analysis above has bypassed the university as a whole since the university is also the dominant organizational infrastructure of the science system and hence a major portion of contemporary science carried out under the heading of *research*. To answer the ongoing question of whether there exists one guiding distinction among the multiplicity of distinctions relevant to the modern university we have to look elsewhere. It is obvious that scientific/nonscientific—the answer suggested by a tradition formerly dominated by the German university model—today only partially qualifies as the guiding distinction. If we want to account, for example, for the fact that American colleges and universities alone have about 300 programs for creative writing,[65] it again becomes clear that scientific/nonscientific is only one distinction among several.

At this point we should perhaps reconsult the rather unsuccessful book by Talcott Parsons and Gerald M. Platt mentioned at the beginning of this essay, for it does promise to offer a theory of *the* American university.[66] What supports this conviction, and can we identify a candidate for one guiding distinction of any given university event? In addition, can we imagine a distinction that convincingly succeeds in re-including what it excludes (i.e., the other side of the distinction)? Even this latter criterion presumably speaks against the guiding distinction between scientific/nonscientific since on the basis of this distinction one cannot explain the fact that even various nonscientific forms of thought and knowledge are nevertheless studied scientifically in the university.

By way of introduction, let me illustrate the special status Parsons assigned to the university with a relatively early quote from the anthology *Theories of Society* (1961). There Parsons speaks of the extraordinary increase in knowledge characteristic of modern societies and continues his argument with the following thesis: "The university system constitutes the main institutionalized *focus of trusteeship* of this great development of secular knowledge and learning. It is perhaps *the most important structural component of modern societies that had no direct counterpart in earlier types of society.*"[67] It becomes clear how far-reaching this claim is when we compare it to Max Weber's similar interpretation of modern society. Whereas the university is mentioned only in passing in Weber's catalog of instantiations of Occidental rationalism,[68] for Parsons it becomes the most important structural innovation of modernity. The reason seems to be that for Parsons, especially in his later works, the theory of rationality coincides with the theory of the university, a coincidence that among other things implies that rationality/nonrationality figures as the guiding distinction for both the university and the theory of rationality.

The fact that Parsons's approach starts from the guiding distinction between rationality and nonrationality[69] sets it apart from most of today's common theories of rationality, which still prefer the rationality/irrationality opposition. This can be studied, for example, in Herbert A. Simon's work,[70] where the implication becomes clear that the entire, gradually obtainable body of knowledge of the mechanisms and laws of human behavioral control and information acquisition is integrated into the concept of rationality such that rationality is on the verge of becoming a concept without a difference.[71] In some philosophical theories, rationality becomes a logical implication of properties (intentions, beliefs) that must be attributed to humans if discourse on humans is to make sense at all.[72] By contrast, Parsons does not include in the concept of rationality qualifications and extensions of our knowledge obtained by way of determining reasons for action. Rather, he moves the richness of further distinctions to the other side, which for him is nonrationality. But once the concept of irrationality, with its negative connotation, no longer figures as the conceptual opposite to rationality, we may become more interested in studying human attributes such as affectivity, expressivity, spontaneity, and informality as forms possibly distinct from rationality, instead of extending the rationality concept unchecked. This interpretation of Parsons's logic of distinguishing also enables us to understand why rationality/nonrationality can figure as the guiding distinction of the university. In the context of the university, the "excluded" side of this distinction, namely nonrationality, signals the thematic inclusiveness of the modern university regarding potential subjects of knowledge; thus derives the fact that there is hardly any form of knowledge or activity that could not be considered, investigated, or taught within the university. In contrast, "rationality" as the side directly defining the university's identity signifies the particular contribution of the university, hence the restriction under which the university places any form of knowledge and activity, or the "surplus value" that it claims to add. In this Parsonian conceptualization we can recognize confidence in the university as the institutional trustee of the Occidental process of rationalization. Once again this stands in sharp contrast to a Weberian position that would rather emphasize the complete discrepancy between the formal rationality of university procedures and the ultimate irrationality of a value decision for the university.[73]

HELMUT WILLKE | **THE CONTINGENCY AND NECESSITY OF THE STATE**

Difference

From the Aristotelian period to the present, the doctrine of state forms has been confined to the question of how systems of rule can be classified. The issue regarding the architecture of public order (be it the *polis*, politics, or the state) has been repeatedly reduced to the inside of a distinction whose outside, depending on one's perspective, has been various things: either one presumed it to be a subordinate internal horizon such as the *oikos*, the family, and, in the modern era, the individual, or one presupposed it to be an external horizon, notably a divine order, religion, and, more recently, society. From the beginning, it is true, the Western tradition has operated with a three-sided form dividing the world into the private, public, and otherworldly spheres, yet only in times of crisis has the quest for an order of rule addressed the other two horizons (the private and the divine), which it otherwise left underexposed.

This asymmetrization of a three-sided form forced onto the question of organizing rulership clearly applies to Hobbes, who mirrors the tripartite form of social order by structuring his *Leviathan* into three parts ("Of Man," "Of Commonwealth," "Of a Christian Commonwealth") but does not utilize the differences. He enlists the side of man and the side of religion only to strengthen his arguments for the state.[1] Similarly, Rousseau makes use of the transcendent (or, in more modern terms, the socially emerging) quality of the *volonté générale* not to reflect the difference between politics and society but to claim a unity between them. In so doing he falls prey to the danger of a totalitarian construction.[2] It was Hegel who for the first time took the form as a tripartite seriously. He revolutionized the reconstruction of rule by relating the trinity of family, (civil) society, and the state to the dialectics, that is, to the many-sided distinction between arbitrariness and rule, between the particular and the universal, between contingency and necessity. As a result, he achieves not unity but difference. The particular altruism of the family is contrasted with the universal altruism of the state, and both are set off against the universal egoism, that is, the unfolded contingency of civil society conceived in economic terms. "In civil society each individual is his own end, and all else means nothing to him."[3] Moreover, this "special" purpose of each person, construed as "a mixture of natural necessity and arbitrariness,"[4] Hegel considers to be one of the two fundamental principles of civil society.

Given the key role assigned to the distinction between arbitrariness and necessity, Hegel can take a completely new view of civil society: It is now conceived as "the difference which intervenes between the family and the state."[5] Marx can then retain the guiding difference between arbitrariness and necessity while viewing society (again understood in economic terms) as the realm of necessity and the state as the expression of an arbitrary system of rule, that is, as a difference that comes between individuals and society and serves as nothing but the normalization of arbitrariness itself.

Only from a systems-theoretical perspective does this proposition lose its outrageousness. In a fully secularized, functionally differentiated society the guiding difference between contingency and necessity no longer separates state and society or politics and economy. It is instead drawn into the operational modes of each nontrivial, complex social system as the path-dependency of a principally contingent evolution and as the expression of self-grounded *form formations* that in principle can no longer be based on external factors. The form of the state, for example, can no longer be celebrated as a decision for or against arbitrariness. It materializes out of the societal function of society's political system as a self-description of politics, that is, as a revisable establishment of temporarily necessary rules for controlling the contingency of political decisions and processes. However, in the course of (Western) social history of the past five hundred years, this was only possible for the form of the state after numerous other possible state forms had had their chance and subsequently yielded both to the force of new ideas and to the power of new social-structural constellations.

In the state's form, the arbitrariness of political processes is condensed into particular patterns for connecting events. Those patterns result from rules whose meta-rule constitutes, in modern continental Europe, the form of the state. We are to understand the state's form as the counterpart to politics freed from traditional bonds. Yet an unrestrained politics would only amount to contingency, thus anarchy and chaos; it was therefore easy for Hobbes in his debate with Bishop Bramhall to argue for the state on the basis of the alternative "Behemoth against Leviathan." The difference between politics and the state allows for the combination of releasing and relinking, of contingency and control of contingency. Hence it allows for that combination of variety and respecification that, for instance, condenses contingent phonemes via semantic and syntactical rules into a language.

If we take seriously George Spencer-Brown's suggestion that we cannot use any indication without drawing a *specific* distinction,[6] then we can infer the intended signified from the underlying difference. The difference within which the state receives its denotation changed historically several times, but later, with Hegel, was expressed in the opposition "state versus society," in which society is assigned to contingency and the state to necessity. In this move the state completes its metamorphosis, evolving from its origin in violence and arbitrariness, from which a territorially overarching alliance of power emerged, to its role as a necessary form of society—a society that in the nineteenth century had already begun to mistrust its own dynamics. Under the umbrella antinomy of contingency and necessity, the state commences its semantic and operational career on the side of arbitrariness and initially leaves to religion in particular the side of necessity. The social history that follows, however, forces a "re-entry"[7] of the difference between necessity and contingency on the side of (secularized) contingency. The reason is that outside of religion and within the realm of politics arbitrariness becomes an everyday occurrence that demands its tribute as well: namely as an undeniable need for rules for stability and hence for contingency control.

In the form of the state, politics creates a guiding idea of itself. After secularization and the awareness of contingency had undermined the external justifications of political agendas, politics had to provide for its own legitimacy. It had to counter the hopeless idea of contingent chance with the secular consolation of formal necessity; thus it had to invent an operational form, namely, a form of state, that overshadows and thereby conceals from direct view the inevitable contingency of any politics. It is therefore quite understandable that up to the present those who observe and feel affected by politics have been declaring this shadow to be the true reality. In this way the precarious combination of chance and necessity can be defused, and the explosively paradoxical self-constitution of the political system is confined to more or less scholarly treatises on the necessity of a form of the state.

Social systems as communicatively and hence symbolically constituted systems are exposed to the full impact of contingency as soon as one observes that social communication is no longer anchored in perennial tradition, religious plans of salvation, or objective truth but is based on self-referentially operating connections of meaning that are alterable and selectable according to criteria the system itself posits. Thus open to observation, social systems too are exposed to debates about alternatives until they eventually have to switch their operational mode from unity to difference, that is, from the continuation of what is always the same to the contingency of identity at a given time. Like psychic systems, social systems need mechanisms for controlling contingency and dissimulating the loss of unity so as to orchestrate this profound transformation. From the viewpoint of social history, these mechanisms differ from each other above all in whether and how they are kept latent or have become manifest in the course of criticism and public education. The latency of politics' distinguishing and self-legitimizing feature—the necessity of contingent rule—could be maintained as long as politics as an order of rule captured its opponent in images of anarchic contingency rooted either in human unpredictability or in the arbitrariness of the gods. This enabled politics to postulate the exigency of its own necessity as the form of current rule.

During the social evolution of a paradoxical enlightenment, the political system liberates itself from justifications based on tradition, religion, or natural law and becomes dependent on grounding its necessity as an arbitrary order in itself, that is, on converting the order of arbitrariness into a function. Like any other nontrivial system, it needs self-description in order to maintain its specific identity vis-à-vis other contingent identities and to organize the innumerable operations necessary for sustaining its identity. Put more simply, the issue is how the system perceives and understands itself in opposition to the world outside it. Once different identities of the political system become possible—for instance monarchy, aristocracy, or democracy—politics is forced to develop a model of itself that supports the contingency of its operations and the operationalization of its contingency. In continental Europe this model crystallized around the notion of the state; however, this was only one of several historically contingent possibilities. In England, the "stateless society par excellence,"[8] the central notion has been "the Crown," and in the United States, the Constitution, with its supporting tradition of the Founding Fathers.

We can now note that on the one hand, forms such as the absolutist state, the liberal (watchdog) state, the constitutional state, the intervening state, and the welfare state define *models of politics* and hence provide the organizational forms for political processes. On the other hand, this political meaning of state forms is also connected to social course settings that as *models of politics' self-legitimization* suggest how a secular order of society at large can be justified. It is true that a secular justification of social order is made possible in principle when a political system is founded on a contract of rule. In practical terms, however, the problem of legitimizing the social order is only shifted to the problem of legitimizing political rule, or more precisely, to the problem of legitimizing the rules by which politics produces binding decisions. From this functional position nearly everything is possible: from the absolute monarch to the general principle of fairness.

Properly understanding the state within modern political systems properly requires a new discourse and theory of the state. This has become more urgent than ever since the state had to lower the veil of latency on a very basic level following the medieval investiture controversy, and even more so after the late-nineteenth-century social reforms and the Keynesian revolution at the beginning of the twentieth century. As a result, the regulating capacity of the state, that is, its controlling of contingency, can be recognized. The theory of the state must readjust to this fact. The experience of modern societies' need to tax has dispelled not only the myth that the state is an autonomous reality transcending social conflicts of interest but also the more recent myth of the state as a mere instrument of the ruling interests.[9] Either myth fails to recognize both the specific function of the political system in a differentiated society and, especially, the operational conditions for self-legitimizing politics. It is politics as a subsystem of a functionally differentiated society that has to carry the burden of the principles and legitimacy of collectively binding decisions. And under secularized conditions it can only perform this task if it focuses on the state's form as a guiding model of its own operations.

As noted above, in continental Europe the *idea of the state* determined the form of politics. In Great Britain the idea of the *crown* played this role, which has meant that the political system up to the present has had to manage without a written constitution, without basic rights, and with tremendous rights for intervention and secrecy on the part of the "government of Her Majesty." In the United States the functional equivalent of the state is the tradition of the Founding Fathers and the Constitution. Despite their differences, these constructions have in common that the political system is based upon a general, "ultimate" legitimizing reason. Without such a foundation, the function of politics—to operationalize arbitrariness—could not be endured. Too naked a view of politics' function would probably endanger the system's ability to make arbitrary decisions and declare those decisions socially binding so as to exclude arbitrariness in the interest of social harmony.

The state, like its functional equivalents, plays the role of the *alter ego* for politics, whose identity enables stabilizing and observing, in a controlled fashion, the contingency of political decisions. Though the idea of the state, the constitution, or the crown changes over time, not everything changes at once. Thus stability can be obtained through the different paces of change. In this way the shock of establishing a secular order and loosening religious legitimacy can be cushioned. In its shift from external reference to self-reference, politics does not engender visible paradoxes but leaves itself the loophole of a guiding idea that, as an internal model of politics, transforms the paradox of legitimacy into a difference of levels. As a result, one can make believe that it is the state that guides politics, and the risky question of who and what instructs the state can be severed from politics, that is, depoliticized.

In a certain sense this construction was already too successful, for it encouraged a semantic super-elevation of the state, especially at moments when diverting attention from the "valleys" of politics became necessary. The reification of the state was a semantic accident of historic proportions. The doctrine of the state is separated from political and social theory. On the one hand, it degenerates into a formal doctrine without having recourse to a proper theory of form, and on the other hand, it flourishes as the juridical discipline of state law by replacing politics as the basis of the state with politics' most certain outcome: laws.

The political system's function and achievement require the execution of physical power to be monopolized, and in civilized, democratic societies, the control of this monopoly must be justified by legal procedures and legalized by laws. Legitimacy, as the hinge between political arbitrariness and legality coded by rights, reemerges as a problem. A solution following the Lockean tradition, that is, the idea of a utilitarian-motivated, consensual social contract, would be to establish the law as the warranty for the contract; a solution in the Hobbesian vein, that is, the idea of a contract of rule based on the fear of civil war, would be to establish the state as the warranty for domestic harmony. Using the formula of the *Rechtsstaat* (the constitutional state), the theory of constitutional law succeeds in uniting the legitimating aspects of both traditions in one semantic construct. Although this naturally has not solved the problem of the legitimacy of legality, the problem is nevertheless sublimated so artfully that the state and the law can be legitimized reciprocally. As a result, the strategically fuzzy concept of reciprocity, as per usual, glosses over the sharp paradox of self-reference.[10]

As a more serious result, politics all but vanished from the agenda of the doctrine of the state. Thus a massive American re-import of sociology and political science was needed after the Second World War to reestablish the political-administrative system of modern societies as an object of scientific inquiry in its own right vis-à-vis the juridical colonization of the doctrine of the state. In 1971, for example, Ernst Forsthoff, in a book that was well-known for some time, could still discuss the state in terms of government and administration, electorate and parties, and even associations without ever making the political system's specific and autonomous function for society an issue.[11]

Through the division of power, the secular political order can become the fully functional equivalent of a traditional, transcendent source of legitimizing social order. This is because within a highly sophisticated and highly improbable architecture of reflexive control and power loops, the problem of distinguishing between arbitrariness and necessity can be organized and eventually transformed into the secular form of the *legitimacy of legality*. With this architecture, the political system with its internal divisions (politics, administration, and electorate) generates legitimacy while the legal system is in charge of reproducing legality. In the simultaneous autonomy and interdependence of the two subsystems, and with the assistance of a built-in self-controlling mechanism, a self-sufficient, functional self-legitimization of politics becomes effective, requiring nothing but two underpinnings. For one, it needs to know what the criteria are for the rules that lend democratic legitimacy to the crucial link between electorate and party politics. Secondly, it needs to solve the problem of how to organize the unity of society if this unity can no longer be represented by the political system itself.

Unity

A completely different form of the modern state surfaces if the form's basis is sought not in the differentiation of politics but in the unity of society, and if the unity of society, though *de facto* no longer existing, is simulated by the form of the state. Political philosophy had centered on the difference between religion and politics during the religious wars of the sixteenth century and on the difference between politics and economy in the course of a dissolving mercantilism. With Hegel, it became possible to conceptualize the unity of the difference between state and society. The tragedy of the Hegelian formula is twofold. First, whereas it posited in the state a unity of the world, François Quesnay, Adam Smith, and others had come to describe the world as comprising economic particularities, and the American and French Revolutions had made it possible to view the world as consisting of political partialities and, in general, as a scenario of the individual quest for happiness. Second, in the notion of civil society the Hegelian formula exposed the sole potential universal—society—to the pressure of a superego that necessarily distorted both sides, the state and society.

Consequently, first Karl Marx and then the entire subsequent theory of the state became trapped in the fallacy of misplaced difference—a history that cannot be unfolded here. Suffice it to say that the difference between state and society was misconstrued for two reasons: first, it suggested reducing the three-sided form of the social world to a duality[12] by dropping out the (private) individual as the capital for heroic worldly projects; and second, because of its underlying notion of the state, it assumed a unity that could easily be misunderstood and misused as a trivially simple whole. In the following I shall outline two manifestations of such a trivialization.

For the German political theory in particular, this trivial holistic conception of the state has left behind a daunting legacy, since it failed to oppose decisively enough, both theoretically and practically, two cynical forms of the state: the SS state and the *Stasi* state.[13] As for the SS state, Eugen Kogon[14] has covered the basics; this still remains to be done for the *Stasi* state. What strikes the sociological observer in both forms of cynical states, though their rationales and effects are quite different, is a deformation that can be described as a threefold hubris: the hubris of simplification, the hubris of organization, and the hubris of power.

These large simplifications not only applied to action, thought, and morality but also included the attempt to reduce the differentiated nature and heterogeneity of modern society to a simple, de-differentiated, unitary hierarchy. "Leader principle" on the one hand and "democratic centralism" on the other were the formulas for an almost panicked suppression of multiplicity. To avoid comparing incomparables, and since sound analyses of the SS state are available, I shall limit my following comments to the *Stasi* state.

Long before Gorbachev and the collapse of real socialism, Peter Sloterdijk had characterized the official socialist discourse as a "cynical speech disturbance of epochal proportions." He continues:

> Even from the outside it is obvious that the politics of the socialist powers no longer holds any hope of socialism whatsoever. In Marxist-Leninist terminology, the East exemplifies naked hegemonic politics. . . . The otherness of socialism has long vanished, but what will happen when it becomes known? Why has the greatest military power in the world been built up in order to protect a fictive otherness?[15]

It has become more clear since then that this cynical perversion of socialism's claims about itself were extended in the same grotesque way to the inside and led the functionaries of the state party to set up a spying, controlling, and repressive apparatus. The *Stasi* not only commanded an inconceivable bounty of human, financial, and organizational resources but also made a mockery of any idea of legality or even legitimacy by engaging in drug and weapons traffic, widespread telephone surveillance, compulsive adoption of children of disfavored parents, training of terrorists and support of terrorist murders, and far-reaching illegal financial transactions and foreign currency dealings.

These internal and external perversions of "socialist" ideas are not accidental outcomes. The fatal flaw lies in the combination of the theoretical naïveté of a de-differentiated society, which is thereby liberated to become a unity, and the practical arrogance of those in power to enforce this vision against reason and opposition. The organization required for the kind of power politics that denies and annihilates contingency deadens all emancipatory aspects of the socialist vision of combining freedom and necessity, contingency and control. The hubris of simplification turns into an organizational hubris that ultimately perverts all of society into a model of the organization.[16] This is inevitable insofar as the forces at work in every aspect of the modernization of society—secularization, increased contingency, functional self-dynamics, and overall heterogenization—counteract the politically motivated attempt to simplify. These forces then can be kept in check only by increased repression, thus by increased organizing of the suppressive apparatus. In the GDR, as in other cases, this was certainly not a one-directional development, but the failure of various "New Economic Policies" proves the self-destructive reluctance of the SED leadership[17] to learn.[18]

Under modern conditions the evolution from the "socialist" state to the SED state to the *Stasi* state did materialize one aspect of the absolute state, a state that since the Hobbesian "Leviathan" topos had been considered a safeguard and guaranty against chaos, anarchy, differentiation, and heterogeneity and had been misused as a means of enforcing unity. The prolonged death of the "socialist" centralized and united state certainly does not prove the necessary failure of any form of a socialist state, but it does forcefully demonstrate the failure of a form of state and politics based on the de-differentiation of modern societies. No modern society can be organized *in opposition to* the unleashing of functional differentiation, the centrifugal dynamics of the unique rationalities of subsystems, and the exploding innovative capacity of decentralized specialization.[19] Merely opposing this sociohistorical trend toward the modern differentiated society leads first to the hubris of simplification, next to organizational hubris, and ultimately, as in the case of the SS state and *Stasi* state, to naked state violence as the last resort for enforcing an unenforceable vision of societal unity.

Irony

What would be an alternative to the model described above?[20] Richard Rorty[21] provides the most compelling and rigorous argument for recognizing contingency under modern conditions as the basis for communication rather than glossing over it either with metaphysics or with forced simplicity. He proposes conceiving of language and worldviews dependent upon language strictly as idiosyncratic vocabularies that despite all changes remain precisely this: contingent semantics that can be superseded by other possible semantics yet that never approximate in any way "true reality" or "ultimate truth." Those who conceive their own language and their own truths in this way yet do not despair or turn metaphysical[22] Rorty calls *ironists*: "I use 'ironist' to name the sort of person who faces up to the contingency of his or her most central beliefs and desires."[23]

The ironist gets herself out of this double bind by taking neither herself nor the world completely seriously, yet taking them seriously enough to work on a new vocabulary that will describe the world for her more appropriately. She is motivated by the conviction that her own vocabulary, too, is tentative and that it can profit from consideration of other contingent vocabularies. Thus for Rorty the ironist is characterized by the development of an "imaginative acquaintance"[24] with alternative contingent vocabularies, that is, a capacity that used to be called empathy. Furthermore, she is able to comprehend many different vocabularies or semantics, thus to orient herself to different language games and discourses and to take seriously the contingency of her own vocabulary. An ironist sees herself as being part of a community that is, though constituted by "collaborating eccentrics,"[25] nonetheless a collaborative community, a community of subjects who respect one another as autonomous and interdependent.

For the private and personal domain, Rorty's proposal exhibits the cheerful air of an elegant rebel: everyone makes her own language; any vocabulary is contingent yet the contingency is endurable through ironic distance. It is somewhat irritating that only literati and professors of philosophy can realize contingency and irony, but then again, lower forms of life are in principle not excluded from these pleasures.

Despite some strikingly weak points in his arguments, Rorty's view is stimulating and pertinent to a theory of the state that takes its departure from the loss of conventional certainties about society's order and the form of the state. Theories of society as well as of the state have to face the fact that precisely what they take for granted is contingent. As the twentieth century comes to a close, they can no longer overlook the fact that, after many centuries of secularization, rationalization, and juridical positivism, the basis for a belief in final forms and in an order transcending time and change has vanished. Not only in a general way but also specifically with respect to their own object of inquiry they must draw consequences from a variety of phenomena: the idea of the constructability of the world; the shift of the temporal perspective to a future that requires shaping; the ubiquity of contingency as the basis for the operations of psychic and social systems; and consequently the issue of the responsibility of people and social systems for their decisions under the condition of contingency.

Interpreted from a difference-theoretical viewpoint, irony has a special way of connecting operations and observations: it de-dramatizes operations contained in forms by handling them as if they were observations. Whereas form-guided operations, insofar as they devalue what they exclude, produce asymmetries and create motives for the reliable connection of what is excluded, observational distance allows the reestablishment of symmetry since no prior preferences are involved. If operations are successfully supervised by observation, then each asymmetrization is enriched by the concurrently supplied possibility of resymmetrization. Consequently the categorical difference between operating and observing, acting and reflecting, and deciding (*qua* distinguishing) and reflecting (*qua* introducing new distinctions) would be sublimated [*aufgehoben*] into an operational form that realizes it is constantly observed and that therefore does not mistake itself for a final point.

The lost security of an order beyond time and change, however, pertain to more than the present conspicuous case of the thoroughly failed socialist experiment. In a similarly profound manner it also applies to the following ideas: that of the final form of a market regulated by capital; that of a logic of science based on the epistemological interests of the natural sciences; that of an educational system founded on discipline and supervision; and that of a health-care system that mercilessly avoids death. Despite their undoubtedly systemic and, in many ways, humane and emancipatory success in liberal, functionally differentiated societies, these ideas are by no means final forms. And if their present implementations make them appear splendid against the backdrop of the collapse of what is allegedly their key alternative—the forms of developed socialist societies—it is precisely this comparison that is misleading. Serious challenges to the guiding ideas and forms of liberal societies emerge exclusively from the inside of those societies. Challenges evolve from alternative blueprints of future reality if those blueprints, even in view of the smashing success of the current forms, dare to point out the dark sides of this success.

The ironist distances herself from her communications and actions because the distinctions that support them can never be final. She can never take the available distinctions completely seriously since tomorrow might bring more reasonable and convincing ones. But she also knows that this lack of commitment and seriousness—this insistence on contingency—is not easy to bear, especially for others. The urge for things final informs the search for a form of the state as it does the search for final forms of knowledge, belief, or love. In dealing with uncertainty and complexity, the systems are always drawn to the side of alleged security and closed standard operating procedures. The categorical imperative of any hermetic system is: "Thou shall not observe!" It takes satanic curiosity to counteract the power of finality with the spur of *intrepid observation*. Not by accident does the devil's career start with a forbidden observation, that is, the observation of the observability of God.[26] Kierkegaard succinctly expressed one consequence: "Just as philosophy begins with doubt, so also a life that may be called human begins with irony."[27]

The state's irony materializes where the state deals with itself ironically. Having said this, we do not mean to say that the state, understood as the self-description of the political system and hence as a model and imaginary entity, takes on an ironic distance from itself or ironically smiles to itself. Rather we mean that the state as a model and imaginary entity *realizes an operational mode* that an observer can describe as ironic if such an observer operates with an appropriately elaborated notion of irony.

KLAUS P. JAPP | **THE FORM OF PROTEST IN THE NEW SOCIAL MOVEMENTS**

Social Movements and Society

I should state in advance that to my knowledge a form-theoretical analysis of the new social movements does not exist, save for initial attempts in some of Niklas Luhmann's works. Given the objective of integrating a theory of the (new) social movements into the general theory of social systems, what makes such an analysis especially appealing is the possibility of coming closer to a general theory of social movements. Naturally, we can furnish only a kind of preliminary framework here, which need not be discouraging, for one has to start somewhere.

First, I shall distinguish between old movements and new movements in order to delimit the "new social movements." The old movements, such as the various labor movements, are tied to questions of normative law and to distribution issues; moreover they work with strategic orientations. New social movements (those that developed after the welfare state, namely the ecology movement, the new women's movement, the peace movements, "autonomous" youth protests, and alternative economic projects), on the other hand, have to manage without socially universal reference points for normative and distribution-related definitions of justice. Functionally differentiated societies do not provide unifying, readily accessible self-descriptions of collective deprivations that could be altered and strategically combated through collective protest. New social movements must generate themselves.[1] Thus one might ask why they have to exist at all. New social movements have, of course, specific founding dispositions, such as the boredom of modern individualism, various arousal potentials of communities, or personal concerns [*Betroffenheit*][2] that can stimulate the emergence of collectives based on the communication of anxiety. For the moment, however, I am interested in the societal side of the "demand" for movement protest. In accord with

Luhmann,[3] my starting point is that (modern) societies tend to organize self-descriptions by way of, among other things, an internal boundary of reflection through *protest communication*. Society sees itself in the mirror of protest, which in turn enacts itself by delimiting itself from society. That is to say, the form of protest—or, more precisely, the form of the social movement's protest—lies in the distinction between social movement and society.[4] The movement's other side is "Society," which, however, remains latent since a strong self-reference leaving everything else undetermined is characteristic of the new social movements. Even the distinction itself remains latent: The "re-entry"[5] of social movement and society—that is, reflection upon society by means of this distinction—results with respect to society in a latency-securing "compression of form"; it leads to the distinction between protest and issue.[6] As was the case for the old movements, society as an object of reflection remains undetermined for the new social movements. Its place is taken by issues that are in a way extracted from the latency domain of society and that actualize the form of protest.

Together these brief preliminary remarks point to the fact that, contrary to common assumption, the new social movements do not enact protest by starting from a contested issue. On the contrary, they first consolidate the "form of protest"—the collective readiness for opposition that can be shared by a community[7]—and *then* they seek issues through which society can be criticized, that is, through which one side of the form of protest can be designated. Self-rationalizations then reverse this course in that they externalize internal reasons for the emergence of protest (protest communication) with the help of issues. One is then prompted by topical issues and does not recognize the triggering function of a predisposition for protest.[8]

What counts as the "reality content" of the new social movements is therefore tied to the observational frames that allow this process to be kept latent. (Self-)observation of the latency of both the distinction and its other side would definitely lead to the erosion of at least *collective* protest. We surmise that strict conditioning[9] against transgressing, or "crossing," primary demarcations and restrictive use of second-order observations maintain the observational capacity of the system "new social movements" for its own specific purposes (and therefore maintain it *differently* for the purposes of others).[10] The other side of the difference—

the unreality of the new social movements (i.e., disintegration of the form of protest, its literal crash into society)—is avoided by limiting the capacity for self- and external observation. The resulting limitations on reflection distance the new social movements from the observational relativism of modern societies, hence their often bemoaned yet hardly understood propensity for bold labels: *the rulers, the system, men.*

We can now differentiate distinctions on two levels: those provided by the social economy of communication (social movements / society) and those introduced by the new social movements themselves for their own systemic setup (protest/issue) so that, among other reasons, they are precluded from seeing the former distinctions. The new social movements observe Society, but they do so through their particular lenses of protest and issue; taking any other perspective would amount to *acknowledging* that one observes society *from within society*, thus in competition with other observations (for example, those of the nuclear-power lobby, of politics, of the economy, or of the media) and without any claim to exclusivity. But what ought not be cannot be.

The Form of Social Movements: A Sketch

The difference between communication and observation may help clarify a phenomenon that has been somewhat ignored in the standard literature on the new social movements, namely the dogmatism[11] that the communication of opposition, once chosen, shows. To explain it one can cite either the difference between those who decide and those who are affected by the decision or the difference between risk and danger.[12] Yet those differences do not explain why they are employed by the social system called "new social movements."

We take up two arguments—one by Luhmann and the other by me. Luhmann's argument runs roughly as follows: Second-order observations (that is, observations by an observer who has lost his innocence) generate contingency; hence "circuit breakers" are needed to switch back to the first level of observation or else a "communication overload" will ensue.[13] This consideration prompts me to reformulate one of my own theses. The issue here is *how* the new social movements[14] actually manage to prevent the crossing over from one side of the form (which we define briefly as communication of opposition) to the other.[15] By contrast, the question of *why* it happens is relatively easy to answer: If reflection in the collective communication of opposition were not stopped, the foundation of that communication would disintegrate, that is, commitment, motivation, and hope of some level of success.[16] That is why the other side must remain an unmarked state that blocks the crossing. Observation is thus restricted to first-order observation on one side of the form only in order to keep the operation of opposition closed. This brings us into the midst of the problem of (observing or rather *not observing*) latency.

This connection is obvious if we consider that the opposition to nuclear power plants has become problematic at least since the dramatization of the ozone hole. The other side of the opposition has become generally acceptable for precisely the same formal reasons that pertain to the marked side. Nonetheless nothing changes. Something similar could be observed in the activities of the peace movement during the Gulf War. Various parties bemoaned the one-sidedness of the anti-American stance without noticing what the peace movement could simply not see, namely that a debate between the two (or more) sides on the reasons for war is part of the business of politics, whereas for the peace movement such an attitude would immediately erode the collective ability to act. And it is precisely this ability that counts!

It is my argument that the new social movements enforce communicative limitations by emphasizing what is expected of the membership whenever the injunction against reflection, mentioned earlier, appears to be threatened.[17] The distinction between pro and con is "controlled" by the distinction between "being part of" and "not being part of." Recursive use of this expectation constitutes and consolidates the system, and its communicative form is established by way of a rigid distinction between "being pro" (which is oneself) and "being con" (which is society and its contested issues).

If this is indeed a case of second-order observations, then they are "frozen" within the system so as to hold the system's external reference *on one side* and hence to make crossing the boundary so arduous that few will attempt it. Anyone whose opinion differs from that expected of true members of a movement either just stays away or reduces himself to a mere sympathizer. The system uses observations that evoke other observations focusing on communicative membership. Both sides of the difference "being part of" / "not being part of" are required to discern, by means of this difference, the aforementioned first-order distinction and thus equip the system with the ability to observe itself. Only one side, however, is recursively designated: membership. The other remains latent because it endangers the system. This then leads to the conjecture, which is worth exploring, that members are not "expelled" but rather "stay away," with the result that a ranking of "hard core" members, fellow travelers, and those merely sympathetic to the course develops.

Only an external observer could use the distinction "being part of" / "not being part of" to observe, in addition, the latency of the other side of the communication of opposition. Internal observers shield themselves against precisely this possibility in the interest of avoiding "communication overload"—a precaution that cannot be attributed to any particular *intent*.

The recursive communication of the conditions for membership stabilizes a frame of expectations whose transgression requires time, argumentative energy, and stamina in the face of social conflict. Surmounting such hurdles of communication is therefore unlikely. More likely is the repeated consent to the prevailing expectations of membership.[18] The boundary remains blocked for the system since all further connectable distinctions are related back to this one distinction separating the system from its environment; or in more precise terms, they are related back to the iterative *one-sided* designation of this very distinction. Second-order observations thereby interrupt *themselves* and enable the reproduction of first-order observations. They make possible the form of *one-sided* opposition.

The new social movements are therefore social systems that condition the crossing over from one side of a central distinction to the other in such a way that crossing becomes unlikely; the price they pay for their collective ability to act is a high degree of rigidity.[19] Such artificial yet "simple" systems are made possible by the way the systems use distinctions and impede the crossing from side to side. The mass media, on the other hand, make crossing easier since the media are expected to exhibit a certain degree of neutrality.

This argument also explains why the new social movements gear their distinctions whenever they can toward possible catastrophic events that have a low probability of occurring yet a high (even unlimited) potential for damage:[20] Because such possible outcomes are not short-term issues, they do not lose their currency; hence they establish long-term commitment and stabilize the difference between protest and issue. Consequently, this difference cannot be drawn arbitrarily. Occasions for re-entry starting from this distinction disappear in the all-consuming potential disaster. Anyone who doubts this does not belong to the movement. At the same time communications of anxiety are consolidated; they reinforce the observation of membership by utilizing the distinction between what one ought to fear and what one must not fear. With this focus they produce a stable *eigen*-value that is off limits, especially for debate. Only in the topic of possible disaster does collective identity situated in the latency of the "blind spot"—that is, of the nonidentical, alien multiplicity—become a protest-communicative form within the medium of social polycontexturality.

The collective identity we mention here is connected to the (common) notion of a collective agent or agency. To avoid a concept of an agent who can be identified only by his "interests" and whose interests in turn can be anchored only by going back to this very "agent," we start from the assumption that "agents" are constituted when systems *describe* themselves successfully as "agents."[21] This presupposes reflexive communication, that is, communication about the system's communication whence the very collective identity of the system can result. In the context of the new social movements, the transformation of anxiety communication into protest communication takes place on this reflexive level. By focusing reflection on the potential for technological-ecological (but also cultural) disaster ("dread risk"),[22] the action system gains its collective identity. To the extent that communications are geared precisely toward this identity of the system, the system is consolidated as a collective agent. And it is quite conceivable that the "members" of the system can get by with only a small "investment in identity,"[23] since the lack of a strategically feasible goal to revolutionize social relationships makes unnecessary the type of substantial investment that is required of, for example, members of the classic labor movement. The new social movements cannot provide their supporters with a worldview for life.[24]

That communications are oriented toward this collective identity is guaranteed by another distinction, one that, as it were, "rounds up" all the distinctions noted thus far: On the one hand, there is the self-reference of the system manifesting itself in a mode of communication that binds participants together in solidarity, or to use stronger terms, there is a demand for unity inherent in the expectations placed upon members. On the other hand, and in sharp contrast to the former, there is society's communication, which is portrayed by the movement as lacking cohesion and therefore dangerous.

Furthermore, societal self-observation is initiated in this way, since society is reassured by expert opinions and their theoretical models,[25] which are based on rules of probability. By contrast, orientation toward disaster *potentials* aims at possibilities for damage beyond all calculation, that is, at the "ecological difference"[26] between society and its environment, including actual people, thus also men and women. It is quite likely that the function systems of society would be overtaxed by this (total) distinction.[27] Thus the question arises as to which ways of intensifying self- and external observations can become standard practice to help reduce this overburdening. Peter Fuchs speaks of "cultures of observation" that could become effective in this context; the section of his book addressing this topic, however, is entitled "Speculation."[28] In addition, the new social movements engender, through both their social indetermination[29] and their internal (over-) determination, a zone of uncertainty for politics which increases the movements' noteworthiness precisely when they attract attention anyway.[30]

Issues of Inclusion and Exclusion

1. I rest my understanding of the conditioning of boundary crossing on Luhmann's definition: "The crossing of the boundary can be conditioned. The form of the distinction is therefore the unity of an (internally regulated) duality."[31] This possibility of conditioning is central, for it enables the observation of differences with regard to, for example, tight versus weak conditioning. Communicative pressures for unity effectuate tight conditioning pertaining not only to the new social movements but also, for example, to professions, especially to expert cultures in the area of risk assessment.[32] Conditioning is the key to the "one-sidedness" of designation. The temporal, material, and social investment required for crossing boundaries can be raised to such a high level that one sticks to the position against nuclear power plants even though they offer ecological advantages that obviously cannot be designated. Naturally, condi-

tioning does not appear out of nowhere. Certain factors are required, such as orientation toward potentials for disaster and expectations placed upon membership that are properly geared toward those potentials. The conditioning of the boundary is thus, in a figure of speech, *extracted from* the latency domain of pre-attuned expectations and therefore remains invisible *as a conditioned boundary*.

2. Construing the opposing side as dangerous, unacceptable, and foreign colors[33] the designated side as a pressure toward unity and acute sensitivity to deviance. This presumably constitutes the gist of the *unintentional* conditioning of the crossing: Whenever a social system is erected, this social system will describe itself in reference to solidarity and morality. Those references lead to self-reinforcement by means of *semantic discourses of unity*[34] that include an acute sensitivity to deviance, which by itself impedes the crossing of the boundary. The opposite side can be "set up" in such a way that its own characteristics—in particular its polycontexturality, its lack of "real" people, and its being a mere system that needs to be re-moralized—provoke a structural preference for the "right" side. Communicative sanctions are neither ordered nor established discursively, but rather result from the (recursively generated) *eigen*-value of solidarity and moral bonds against a hostile side that is, apart from its moral unboundedness, not marked further. Communicative limitations therefore remain latent and do not interfere with the continuity of protesting. All this would be impossible if the other side were marked, because communicative restrictions would then become *manifest*. All else aside, these insights allow us to explain why the new social movements possess neither a theory of society nor a theory about themselves.[35] The system's self-reference is formed by way of solidarity, whose other side consists of the *abysmal* (nondesignatable) *ruthlessness* of power and money. Such an "opponent" cannot be fought *strategically*, but can only be rejected *morally*. This difference ties in with the new social movements' strong orientation toward shared experiences, for example, collectively braving the superior might of water cannons. By contrast, strategic *actions* for a specifically different future lead a shadowy existence.

3. The protest-*form*ing distinction (pro/con) remains latent since the distinction "being a member of" / "not being a member of" is attached to just one side of it. The latency[36] can only be observed by an external observer (deviant members, function systems, or thoughts of some con-

sciousness not communicated). Two additional insights can be appended here. For one, it seems as if psychic systems serve, according to the slogan "thoughts are free," as an outlet for the communicatively hermetic nature of the new social movements. Secondly, the new social movements redirect doubts expressed about their extrasocietal position toward the distinction between the ideal society (they themselves) and everything else.[37] Finally, within this context the notion "ecosystem" most likely acquires meaning; to the extent that this notion does not permit a boundary any longer, it allows imagining an ultimate harmonic "unity." Or to express the point differently, it serves as the imaginary unity of the difference between social movement and society—imaginary because in no longer permitting distinctions it cannot be marked. We encounter here a utopian blocking-notion similar to the former ideal of communism: Nothing appears to be worth distinguishing, yet such a position contains the (again latent) distinction between what is and what is not worth distinguishing. It would in effect amount to the "top paradox" of a fundamentalist-minded observation of "Society." Any attempt to *form* monopolistic observations of "Society" ends in an idyllic and totalizing view. It would certainly be interesting to scrutinize from this vantage point the intolerance of many nonsmokers, radical cyclists, and, naturally, members of the German Republican party (*Republikaner*) and other very first-order observers of totality.

4. The distribution of concern through the difference between decision makers and those who feel affected by the decisions generates structural dissent. Moralizing this dissent serves to limit the dissent through specific attributions.[38] Here especially the focus on blame (thus on the culprits)[39] for potential disasters[40] takes effect, leading to selections of those issues that can channel scattered communication of protest into the formation of a system.[41] The selected issues can involve neither everyday, highly probable risks (e.g., those associated with bicycling or smoking), which accordingly cannot command attention for long, nor moderately probable risks on which the mass media are focusing (e.g., toxic substances, whose negative impact on health is controversial). For society's communication media something similar may apply, although for different reasons. Power-oriented communication in the political system, for instance, is scarcely irritated by everyday risks or by those with the potential for disaster.[42] The former attract too little, the latter too much at-

tention. However, risks with a moderate probability (that is, accidents but not worst-case scenarios, such as meltdowns) allow for connectivity in the political system and fall into the domain of what is feasible. For that reason alone, dissent from the communications of the function systems, which is essential to the continuous existence of the new social movements, cannot be dispelled. Conversely, this does not mean that the function systems never listen. When they listen, however, they translate fear of catastrophe into communication that, in the domain of what is technologically, legally, economically, and politically feasible, demonstrates the *state's* ability to act. It is clear that this will always be too little for the new social movements and just leads to extending the limits of possible communication of protest further and further. Other observers (who could be us) might detect in this behavior of the function system nothing but a normal "immune reaction" of society.[43]

Be that as it may, only low-probability potentials for unlimited damage (above all, technological worse-case scenarios, but also war, whose catastrophic escalation can never be precluded)[44] lead to a kind of permanent attention with attendant cohesive effects within the system of the new social movements. Complementarily, the specific forms of communication of protest imprint themselves into the medium of communication, namely by means of *protest contingencies*[45] that are promoted by the socially pervasive difference between decision makers and those affected by the decisions. Through continually new, concern-engendering chains of decisions, society offers a continuous sequence of occasions for protest without itself becoming the object of protest, at least not in a *manifest* way.

5. Being concerned denotes the disruption of personal conditions that one considers important. To put the matter generally, being concerned increases one's individual options (in the labor market, the lifestyle market, and the marriage market) yet simultaneously depletes collective ligatures (i.e., limitations of contingency by tradition, such as social rank and class).[46] Thus being concerned increases risk by increasing options without the accompanying security of collectivity. Thus concern places individualism in distress. Historically, new collectivities are constituted through reactions to chains of decisions imposed by others whom one does not know at all. Against the backdrop of the amoral poly-contexturality of risks and dangers, the move to insert concern into unitary forms

of solidarity and morality becomes comprehensible.[47] The unfamiliarity of the poly-contextual state of confusion is countered by the collective unity of self-reference based on solidarity and morality. This collective unity serves as compensation for the temporary disruption of "one's personal conditions."

6. The interrelation of solidarity and exclusion is constitutive for the new social movements. Unity through solidarity is generated with self-descriptions constructed by means of do's (the semantic discourse of solidarity) and don'ts (dissociating reflections) for communication. These guidelines lead to the conditioning of the boundary between conformity and deviation. Already the designation of the *unity* highly sensitizes the members of the group against deviation, that is, against changing sides. This relatively simple insight makes clear why form-theoretical observation of the new social movements offers advantages over traditional approaches. The poly-contextual pressure toward communicative unity is nothing but another manifestation of a basic societal tendency to react to excessive complexity by retreating into *self*-reference. And self-reference always presupposes "the other side" that we are constantly mentioning, that is, external reference. It entails a distinction, either side of which could be marked for itself even though there is no reason—at least for the new social movements—to *know* that one makes such a distinction.

7. The formation of unity on the designated side (that is, on the side of the social movement) inherently makes latent the fact that latency (of the other side) does not matter. To that end the new social movements need precisely the conditioning of the crossing by means of the do's and don'ts of communication we mentioned earlier. If the other side (in general: what is not unitary, the poly-contexturality, the complexity of risk trade-offs) were to be observed as *latent*, then the "internal regulation" through conditioning of the boundary would cease to function. Moralizing collective identity serves as a protection against latency. This construction makes the (Habermasian) imperative of universalizability of reasons superfluous since one's own moral superiority always presumes the universalizability of one's own reasons without the necessity of subjecting those reasons to any "rational discourse." The new social movements function in this regard like social institutions: participants know what is meant without having to discuss it first.

8. The new social movements must negate the relevance of latency

(the other side) so they can observe the latencies of the function systems in a "unitary way" (i.e., the *system-relative* heterogeneity of the "ecological difference" between *society and its environment*). They do so paradoxically in view of possible disasters, without realizing the constructed nature of this procedure. Anything else would amount to the self-dissolution of the collective communication of protest by directing the focus to the other side, that is, the manifestation of latency.[48] The preference for collective experience also serves this mechanism, and such experience still remains, paradoxically, individual experience when one acts collectively. For the communication of protest is always about *protest* and not about some strategically calculated (intermediate) steps toward a visionary goal.

9. Second-order observations lead to resymmetrizations. Both distinctions (acceptance/rejection and "being a member of" / "not being member of") would fail to bring about a complete resymmetrization of their two sides owing to the biasing effect of what is distinguished. They would, however, irritate the conditioned asymmetry because the latency of the other side would be disclosed. Since re-entries cannot be avoided within the evolution of a system, they will flare up briefly because of the associated communicative prohibitions and be absorbed by the system's built-in drive toward unity. Someone will ask questions that the rest of the group already knows do not belong. One either does not listen to or does not contest the significance of the questioning. This is not in the least due to the re-entry that brings the system into its form: The self-mirroring of the distinction between social movement and society pushes off "Society" as something foreign (as the "bottomless" multitude) into the unmarked state of the other side. And this goes along with the corresponding solidifying effects of the designated side, which is reduced to the form of protest and issue.[49]

10. Re-entries will only uncover the latencies of the system if the difference between protest and issue disintegrates, that is, when the system's external reference must be newly constituted. Then conditionings of the boundary dissolve, and the system enters a state of imbalance.[50] Yet this is precisely not the everyday reality of the new social movements. Their everyday existence consists in *not observing* the latency that renders them possible in the first place. They cast their non-everyday reality into the latency from which they regularly bring back the continually erupting difference between decision makers and those affected by the decisions into

the manifest state.[51] This again points to the fact that the form of protest is a sort of permanent social fixture securing its hold in "suitable" issues and not in "Society," though society makes it possible.

The Observation of Society

All the above (in every way preliminary) considerations evoke the question of which forms "Society" mobilizes to observe the new social movements. On this level we can distinguish three approaches.

From the viewpoint of society's *perception of risk*, the differentiation of center and periphery comes to mind.[52] From the center the social movements are subjected to distinctions such as normal/deviant and rational/irrational (especially in relation to expert cultures), and what is considered normal or rational remains latent in each case. This results from leaving the basis for consensus unexamined, an omission that first and foremost enables designating one side of the distinction without reflecting on the unity of the distinction. The "innocent culture" unfolds its effect, and the functional position for societal self-observation remains vacant.

From the viewpoint of the risk-averse *mobilization of resources*[53] (take for example the youth movements of the political parties and churches), distinctions such as co-optable/non-co-optable and politically risky / politically negligible become pertinent. In this regard social movements are more or less integrated into the process of social change, although these are *organized* movements ("movement organizations"), which at times attain the power of formal organizations. They are thus not like new social movements that are autopoietic,[54] but rather belong to the "center" in their orientation. Hence, when observing this kind of movement the center observes only itself without distinguishing this "self." Consequently, the functional position continues to remain vacant.

Things become more interesting when we focus on the "archetype" of the new social movements, including all potential and actual effects of blocking the political and economic production of decisions.[55] In this context implementation deficits arise especially in state decision-making bodies that deal with waste-disposal facilities (toxic, nuclear, and otherwise) perceived as hazardous. In these cases observational forms do not result primarily from forms of risk perception, or from distinctions related to a movement's mobilization of resources. Rather they are based (if violence is excluded) on *risk communication*,[56] the function of which is to negotiate understanding; therefore distinctions completely different from the ones just introduced are required.

Nonetheless, we can learn from both the case of society's risk perception and that of the movement's organization. The rational/irrational distinction, for instance, suggests communicative operations that count on information, on public education, or simply on manipulative rhetoric of conviction. It has become common knowledge that such stratagems are counterproductive for risk communication since they intensify the opposition they set out to reduce. One could have known this all along if the latency of one's own (that is, the "rational") side had been observed along with the other side.[57] One's own rationality (or "innocent culture") could then have been regarded as contingent, and the improbability of the new social movements' taking on foreign (and dubious) assumptions about rationality would have been registered in the risk communication. Yet this is just another case in which the rule applies that one only sees what one sees and not what one does not see.

Where movement organizations are concerned, distinctions are relevant that suggest co-optative strategies. The perception of danger is mitigated by buying off those who are concerned with participation in decision making.[58] Participation in decision making, one could also say, leads to neutralizing counterobservations (of the side of those who are concerned) and thus allows the latency of the official risk preferences to remain latent. Nevertheless a reflexive retardation of the pace at which technological-ecological decisions are reached could gradually assume normalcy and thereby increase attention to accidents and heighten sensitivity to surprises. S. G. Hadden demonstrates these effects with the case of a planned toxic disposal facility, the construction of which was fiercely contested at first but came to be accepted via successful co-optative strategies associated with retardation.

Yet with regard to the new social movements being described here, all this does not amount to much. Aside from *specific* reactions of concern (to the so-called facility siting) the movements in question also display a *general* decrease in willingness to comply with social institutions, an attitude manifested in universalized distrust of institutional elites, political parties, and more broadly "the other side" (i.e., "Society").[59] This brings us back to the paradox of the "ecological difference." For the new social movements, this difference implodes in the envisioned potential catastrophe, with the result that the difference itself disappears in the *imaginary* unity of the catastrophe. Function systems of society can interpolate the difference between risk and danger, and each can describe, from the perspective of either risk or danger (depending on their view of the problem),[60] their *specific* unity of the ecological difference. The societal unity of that difference is not attainable; it can indeed only be imagined as a designation of one side of several or many second- or third-order distinctions (vis-à-vis the primary ecological difference) that aims in each case at a specific unity.[61] These distinctions and their significations can only come about in the context of poly-contextural possibilities for observation. Yet it can be assumed nonetheless that the social network of observations guided by the ecological difference is *intensifiable*.

In this regard we see possibilities for resolving risk-communicative aporias and, by extension, for the protest-communicative self-description of "Society." That such possibilities exist is substantiated, for example, by the communicative resonance of protested issues in the function systems of society and by the increase of intermediary systems of negotiation (especially on the municipal level) resulting from, among other things, the *structure-forming* effects of those conflicts discussed above in the context of function systems and new social movements. These negotiation systems have the purpose of transforming *prima facie* insoluble conflicts (characterized by strict nonobservation of latencies) into solvable conflicts (characterized by loose nonobservation of latencies).[62] They do so by enabling provisional agreements in risk communication.[63] Those agreements probably only come about if the conditionings of boundaries (not only of the new social movements) can be loosened without undermining the required protection of latency. This constitutes a veritable paradox since the latency protection consists in appropriate conditionings of boundaries. This paradox may only be dissolved through temporalization, thus by deferring on currently irreconcilable issues. But this is a separate topic.[64]

One can certainly assume that the chances for evolutionary survival of a risk society are increased by the reduction of the difference between technological-ecological obstacles to implementation (of incineration plants, for example) on the one hand and agreements (on these plants), however provisional, on the other. At least, this holds if one shares the premise that societies that increase their difference from the natural environment without check will be eliminated by evolution.

GIANCARLO CORSI | **THE DARK SIDE OF A CAREER**

I

The possibility of building a future for oneself by means of one's own actions is not an exclusively modern phenomenon. Even in the late Middle Ages the individual could be observed and judged from the viewpoint of his or her life history, at least in some social areas; with regard to spiritual salvation, for instance, the entire individual biography was deemed relevant and not only the act of repentance during confession on the death bed.[1] As long as society is stratified, however, only the upper stratum can function as a model for moral behavior since only members of this stratum have destinies of their own, determined by their lineage. For them there is no difference between a destiny deserved and one not deserved, between merits attributable and merits nonattributable. Others can follow the example. If one wants to acknowledge and acquire merits, one can learn how to do so—but it will remain discernible that one had to learn first.[2]

Lineage and career are very different ways of determining the individual's future. Time in the lineage paradigm is bound to a given order against the backdrop of divine eternity. The upper stratum can realize its own destiny without depending upon the contingencies of life. If one, however, does not belong to this stratum and cannot count on social qualities of one's own to make oneself observable, one must resort to some form of planning and shaping one's future. It is likely that determining one's future initially became a problem for those who could not give their lives meaning in reference to God's will and for whom neither chance nor necessity were acceptable as determinants of their future.

The late-medieval merchant is an example for such a social group. "Merchant time" was not "Church time": It was a newly introduced temporal dimension, one that required a certain degree of planning of one's activity and that continually forced self-attribution of what one had done and had not done. The difference between the two temporal perceptions becomes clear when we consider the Church's stance on the matter of usury. For the Church, practicing usury meant trading in "the time" of others, thus time that belonged exclusively to God.[3] In the Christian framework, "time" signified the individual life itself and not simply future opportunities: Life is a gift from God. How could one permit the given destiny of an individual to be codetermined by another person?

The fact that the history of a person presents the possibility of establishing that person's quality forces society to come up with a different idea of what a person is. In a certain sense, a true "discovery of the human" occurred in the Renaissance.[4] The life of the individual was no longer exclusively determined by birth, family, or social stratum; rather, apart from innate qualities of the person, it can be changed during one's lifetime. It is not by accident that merchants and not aristocrats were the first examples of this new development; after all, how could an aristocrat forgo the solidarity of his or her own stratum? For the members of the upper stratum the appeal to their dignity sufficed to make their respective futures visible. But the other strata began to imagine the individual life as a series of stages. And this led to a new structure: the career.[5]

Today lineage no longer plays a role for the destiny of the individual, and we all equally face the problem of managing one's life. Lineage and career are no longer alternatives. Education is the best example of this; it demonstrates very clearly how the perspectives have changed because it is a social system that constantly deals with the issue of determining the future of personal lives. A typical problem addressed by the pedagogical discussions of the last decade pertains to the design of schools: Can we assume that the future is open and hence will be codetermined by the pupils' interests, or should we fix from the outset the curricula that are to be followed? The problem would not arise at all if it were still possible to entrust oneself to a destiny predetermined by birth.[6]

The relationship between the individual and society seems to have taken on a new form, and the career is this form. The future of the individual is no longer shaped by his or her position within a given order but is determined by a chain of events that acquires a certain logic only from within.

II

The concept of career does not designate only the positive side of a course of life, that is, what is usually referred to as "making a career for oneself." Even if one's biography takes a negative turn, an individual history is constructed nevertheless since negatively evaluated events also influence and delimit what can happen next. Hence the negative has to be included in the concept as well. The concept must also cover cases in which someone decides not to make a career, for here, too, one is exposed to the consequences of that personal decision. Whatever decision is made, it can be observed, and it contributes to the formation of the person's social identity.

If we refrain from defining "career" on the basis of quality, we can conceive of it as a digitizing of the continuum of an individual biography wherein the single digits represent events that have the following characteristics:[7]

1. Career is not a structure of the individual consciousness but rather a social structure. This is evidenced by the fact, for instance, that career-building events are always a combination of selections made both by the career subject and by others. It does not suffice that the individual alone makes a decision; others must have the possibility of deciding positively or negatively. One can apply for a position and be accepted or not. One can propose marriage to someone, but obviously a complementary decision is needed as well. To define "career" we therefore need take into account the distinction between ego and alter ego, that is, between self-selection and external selection. This makes the concept of career one that is determined as a social structure.

2. Any event delimits and thereby makes anticipatable what can happen in the future. At the same time, however, the career is by nature contingent. On the one hand, the future is uncertain, and it is hard to predict what will happen even if possibilities are utilized for structuring time by means of, for instance, high school and college degrees, professional diplomas, and so forth. On the other hand, one does not know whether the past that is continuously produced will be a good preparation for the future. The present continuously puts the career under pressure since only in the present can a past be produced that one will need later.

3. Since the career is subject to contingency, each event is evaluated only in terms of the indications it provides about what can be expected in

the future. The relevant distinction is the one between positive and negative, and the significance of events for a career depends on the direction the career takes. Not everything that happens in the life of a person contributes to the formation of a career, but only those moments that give the course of events a direction, negative or positive.

The career makes an ego observable to an alter ego, especially over time. With regard to a career, it is possible to develop expectations vis-à-vis others, and those expectations contribute to the formation of an individual's social identity. In modern societies the career is one of the most important means of exposing oneself to the observation of others and thereby of functioning as a communication partner.[8]

III

In a temporal sense, the career is a complex structure. It evolves only when it continuously projects a space of possibilities from a given situation, already determined by past ones, into the future. In this operation, selections can only be made in the present, and only in the present can reality be observed as the difference between past and future. A career evolves by excluding and inhibiting potential alternatives. In a certain sense, as what it excludes becomes more clear, the career gains a unique, more specific and improbable identity.[9] From this perspective, the career is the unity of the difference between actualized and nonactualized possibilities—a difference that can only be made from within the career. This feature of the career allows us to speak of morphogenesis. This notion denotes the formation of an irreversible structural history that proceeds from the difference between implemented and excluded alternatives, provided that the difference can be applied only retrospectively and that the morphogenic process develops blindly.[10] The blindness of the morphogenesis is the precondition for morphogenesis itself, a fact that implies the impossibility of determining in advance the direction the formation of a structure (in our case, the career formation) will take. Yet the present also produces uncertainties since future possibilities are present uncertainties: One indeed cannot know what will happen next.

The morphogenesis therefore presupposes an identity (a form), which remains the same even if (and precisely when) the other side of the distinction (that is, what the career is not or is only hypothetically) changes.

In this sense the career denotes the building of an individual identity in the irreversibility of time, where irreversibility means the exclusion of possibilities. The difference between what is chosen and what remains excluded is drawn anew in each instance such that in every career event the already determined past and the probable future become relevant again. The career reappears in itself—a sort of "re-entry" of the form within the form:[11] In order to apply for a position, one must present one's history and intentions. The chain of prior selections constitutes the criterion for actualizing the next selection.

The excluded alternatives remain observable, for example, as paths now closed that were once open because of certain qualities of the career subject. The more complex society becomes, the more numerous are the opportunities for selection. Along with these selections, the awareness about the selectivity of one's own decisions heightens, with the result that the exclusion of possibilities puts increasing strain upon the career.

IV

So that careers can be built, society must provide a large number of social positions that can orient self- and external selections.[12] The organization of roles in modern societies, ranging from job positions to complementary roles (e.g., teacher/student, physician/patient) to positions within the family, may fulfill a function of that kind: Social positions have the characteristic that they can be occupied by several individuals (sequentially, of course) and that they do not disappear when they are occupied. This makes it possible to regenerate options for selection time and again. The improbability and variety of careers is connected to the degree of differentiation among social positions, and the importance of formal organization for contemporary society may be tied to the need to structure increasingly complex careers.

Thus, social positions are relevant not only to the individual career but also to the relationships among different careers.

The career is a structure that provides two possibilities for dealing with complexity. On the one hand, the career is based on a projection of possibilities into the temporal dimension resulting in the creation of a past that makes us expect a future. On the other, determining what is possible requires that others, too, can make decisions about the career subject's decisions and thus contribute to the formation of expectations that structure the career. The relationship between the temporal dimension and the social dimension therefore is one of mutual delimiting and determining of what is possible. If someone decides to aim for a certain social position (for example, a job position), and he or she is accepted, then this has certain positive effects on the career in question. At the same time, however, the possibility is negated for others to achieve the same position; thus a negative event is produced in their careers. For example, if there are too many sick people, the number of patients admitted to a hospital must be limited; if there are too few kindergartens, not all children can be accepted. Deciding to accept someone also implies deciding not accepting someone else. Thus a single external selection can have different meanings and consequences for many self-selections and the corresponding careers.[13]

Even though is it not a zero-sum matter, the making of career events always affects others as well, and attempts to structure one's own time always have consequences for corresponding attempts of other people.[14]

V

Only if we know something about the form of the career can we acquire a precise notion of career. Since "form" itself is a concept, we need to define that concept first. What characterizes the form of form?

In recent developments in systems theory, the concept of form is defined in reference to George Spencer-Brown's calculus.[15] Form is conceived of as "a cut through the world resulting in the existence of two sides. . . . In any instance form is a two-sided form, a caesura, a wound of the previously indicated world."[16] Regardless of what one observes, the observation consists in an indication within a distinction: There is no other way of processing information. The quality of a concept therefore is tied to the distinction out of which it is formed and of which it is one of the two sides: This is the form of a concept. In this sense a concept becomes more precise the more it enables us to determine from what it is distinguished, thus what it negates.[17]

There are, however, special cases where the concept's meaning does not allow for its being distinguished from something else. This is true of concepts without differences; they are particular in that they include their own negation. The concept of the world, for instance, on which the idea of form is grounded cannot be expressed as a difference since the world is precisely the backdrop of all distinctions; it is the unity of any difference and therefore cannot be distinguished.

The notion of the career certainly does not possess the status of the world concept. If, however, it is pushed to its extremes such that the zero career and negative career are still considered careers, one can see clearly that this concept does not allow for alternatives and also includes its own negation. Whatever one decides in a career, one cannot avoid being observed as a career subject: Not only is the negative career a career, but even a person who decides to forgo any career cannot prevent others from observing him or her as someone who made that decision and not another. Even the person who does not depend on external selections has a past and a future that determine his or her social meaning.[18] Thus the conceptual opposite to "career" does not consist in a socially available alternative.

Instead we can ask whether there is any condition of possibility of a career that figures as a dark side of sorts without presenting itself as an alternative. Aging seems to be one such possibility.[19] It is common to measure the improbability of, or interest in, a career by measuring it against the age of the career subject.[20] Moreover, the expectations that can be directed toward a person vary quantitatively and qualitatively with his or her age. For example, they decrease at an advanced age, though this does not imply that biological age and psychological age coincide.

The role of aging is especially apparent in the fact that aging always implies a decrease of career possibilities: One can begin fewer new things; certain options are precluded; the horizon of what is to come narrows; and so forth. Occupation offers a typical example: retirement terminates the practice of a profession completely, independently of one's prior career. In the social politics of the welfare state, we note how, with regard to retirees, attention is focused on the relationships between the temporal and the social dimension, that is, on problems of loneliness, boredom, and so forth. The future remains available as leisure time, as the quest for happiness, as contemplation of what others, especially relatives, do. The contingency that constitutes careers decreases, and its place is taken by its opposites: necessity and impossibility. Contingency can decrease for many, more or less arbitrary, reasons. However, there is no alternative to aging. The extension of the future's horizon becomes increasingly narrow.

Aging entails a kind of "cooling-out" effect on the demands placed on the career: One tends gradually to become content with what one has achieved.[21] One can no longer articulate a future and therefore lessens self- and external expectations.

Age and aging hence seem to be external conditions for careers, thus a kind of pointer that forces one to observe in an analog fashion what is digitized by the career.[22]

The difference between career and aging exhibits another feature that has to do with the way certain distinctions are employed. Distinctions that have a high structural value[23] normally display an internal asymmetry: One side guarantees the connectivity of operations while the other reflects on the contingency of the former side. We can also apply this difference between the value of connectivity and the value of reflection to our case. Age is a good indicator for evaluating a career: Certain achievements or positions attained can be more or less surprising depending on the age of the career subject.

Where previously lineage determined the direction of one's destiny, today it is the difference between age and career that projects the horizon of what is possible, and it does so without a predetermined direction, that is, in a flexible way.[24]

The function of the career is to make the future predictable. It makes possible asymmetrizing the world along the temporal dimension and thereby letting expectations arise which develop beyond the immediate present (with its chance events) and which can be improbable to varying degrees.[25] The career is free, yet not in the sense of "anything goes" (for instance, self-selection), but rather in the sense that the career itself creates its own opportunities and limits, its own irreversible history and uncertainty. This applies to all careers, which in this aspect are positioned on the same level. The individual can observe himself as something constant against the backdrop of fleeting time: The presence of a temporal threshold produces inactuality (i.e., new as well as excluded possibilities); in fact, it engenders a different inactuality for each individual.

Viewed from this angle, aging takes on an important role. Its characteristic as the condition of possibility for careers is based on the fact that society can observe all individuals as being the same even though they are not. Age permits classifying age groups and within those age groups the homogenization of individuals, especially with reference to certain important stages such as high school, college, in the work force, retirement, and so forth.[26] If we bear in mind the initial problem of building a determinable future, we note that the reference to age makes sense only when career asymmetrizes: Being twenty years old does not yet say anything about the status of one's life, whereas in reference to a certain career being twenty or thirty years of age makes a difference.

FRITZ B. SIMON | **THE OTHER SIDE OF ILLNESS**

The distinction between *sick* and *healthy* may be one of the oldest schemata of difference guiding human self-observation and external observation. Likewise, therapy as an operational mode designed to promote the transition from a state designated as *sick* or *nonhealthy* to a state termed *healthy* or *nonsick* is probably one of the oldest social form creations.

Defining form (in Spencer-Brown's sense)[1] as the outcome of distinctions by which a marked area within a boundary is separated from an unmarked area outside of this boundary leads to the question as to what the guiding differences of the distinction between therapy and nontherapy are: what matters are included and excluded by this distinction; which units are constructed?

In answering this question, I shall discern two domains and investigate their correlation, namely that of the observed biological, psychic, and social phenomena and the linguistic domain within which those phenomena are described and explained.

When first considering the *sick/nonsick* distinction, the fundamental problem of such an analysis becomes apparent as one that is located on the linguistic plane. The notion *sick*, as it is used today, is far more than just a denotation of a state. It implies a valuation and an explanation of this state; that is to say, it contains connotative statements about tacitly presupposed evaluative criteria, and it explains the genesis of the signified phenomena by means of mechanisms that are not directly observable.

To separate these planes, I shall distinguish in what follows between describing and explaining phenomena. By "explanation" I mean a "generative mechanism" that the observer constructs to account for the genesis of a described phenomenon.[2]

Although there are numerous findings that permit conclusions about diseases and injuries (phenomenal level) in prehistorical times, they warrant no conclusions (or only very speculative ones) about their explanations and the courses of treatment derived from those explanations. Medical historians usually interpret these findings by analogy to the so-called "magical-animistic" tribal medicine that ethnophysicians have described since the beginning of colonialism.

The medicine of the advanced civilizations of early Babylonia and ancient Egypt provides the first sources warranting statements about therapeutic concepts and therapeutic procedures. The development of medicine in Greek and Roman antiquity can be viewed as a continuation of this ancient tradition and determined the European conceptions of illness and recovery until the emergence of the natural sciences.[3]

When considered from the viewpoint of a theory of observation, the historical development of medicine I will outline hereafter demonstrates that one form, created according to the two closely linked, guiding differences observable/unobservable and comprehensible/incomprehensible (or not yet comprehensible), has underlain all disease models and, consequently, all therapeutic concepts and procedures up to the present.

Likewise the roles of and the relationships between patient and therapist are constituted by the difference in their access, as observers, to different phenomenal domains and in a different understanding of the way they function. In addition, both patient and therapist possess, as actors, different options for action and treatment; the options for acting upon oneself are limited (scratching one's back is difficult enough, let alone transplanting one's heart), and certain operations require, for quite practical reasons, the separation of subject and object of action/treatment.

Symptoms as Signs

To become a patient, one must have symptoms.[4] Only one who displays symptoms is labeled *patient*.[5] Having symptoms constitutes the distinguishing mark by which the observer or observers (be it the patient himself or others) distinguish the patient from the nonpatient.

Symptoms can be defined as observable events, processes, or states that are interpreted as signs for other, *un*observable events, processes, or states located in a second, nontransparent phenomenal domain.

In observing humans, perceivable phenomena (for example, red facial spots, a limp, cries of pain, unclear speech, convulsions, and so forth) are discerned and assessed as states or behavior that deviate from those we usually expect and take for granted (that is, without asking for an explanation). Those phenomena are called *symptoms*, whose meanings an observer or observers cannot construe directly from the communicative context. That is to say, symptoms are not immediately comprehensible; they lie outside of the communication game; they cannot be connected on the basis of rules of communication; they require explanation.

Since no immediate communicative quality can be assigned to symptoms within communication between people, they are interpreted as signs that refer to another, "causal" deviation (distinction)[6] beyond the context of direct interaction and communication, that is, outside the boundaries of the social system.

In most conceptions of illness, this other, not directly observable domain is the body, meaning that a deviation within the boundaries of the body is assumed. However, it need not be the body, for instance, when the psyche is considered to be ill. This state or process, as distinguished from a "self-evident" state or process, is called *illness*. It is construed as a structural and functional change (with respect to only vaguely defined *healthy* functions and structures) within a domain that lies outside the comprehensible social system.

In so-called "magical-animistic" medicine, changes within the body are conceptualized as "embodiments of an illness-inflicting demon."[7] Anyone incensing the demons will be taken over by them; the person becomes "possessed" and sick.

By contrast, ancient Greek and Roman (i.e., Hippocratic) medicine conceived of illness as disorder in the harmony of the four fundamental elements (fire, water, air, earth) within the human body. Galen[8] systematized in his medical works the pathology of humors, which constituted the guiding concept for the art of healing until early modern times: illness, it was thought, is caused by a "dyscrasia," that is, an imbalanced, bad mixture of the four key fluids, or humors (blood, phlegm, yellow bile, and black bile). Thus humoral pathology also connected symptoms causally to intracorporeal changes.

For millennia, knowledge of structures and processes lying below the surface of the skin was quite limited. Though dissections were carried out occasionally, they only served to confirm the doctrines of humoral pathology. At the end of the eighteenth century, when autopsies started to be conducted in the search for pathological structures, it became possible for the first time to connect retrospectively the occurrence of symptoms (during life) with intracorporeal changes identifiable posthumously. In this respect, Giovanni Battista Morgagni's work *De sedibus, et causis morborum per anatomen indagatis libri quinque* (Venice, 1761) was pioneering; in it he directed the physician's attention to visible changes of corporeal structures.[9] The "birth of the clinic,"[10] that is, the orientation toward the body's interior, had been inaugurated.

The preliminary result of this surface-transcending focus of the diagnostic gaze has been the development of numerous technical procedures (ranging from radiology and endoscopy to positron-emission tomography) aimed at making internal structures of the body visible.

More recent systems-theoretical conceptualizations regard organism, psyche, and social system as autonomous, operationally closed systems, with each being the environment for the others.[11] On the basis of those concepts, these events, processes, or states designated as symptoms are, in fact, elements not of the physical system but of its environments (i.e., the social system and/or the psyche).

When symptoms are perceived within the frame of self-observation, we are dealing with psychic operations (for example: experiencing pain, a disordered state of mind, feeling ill, etc.). The phenomenal domain within which they are distinguished is the psychic system. On the other hand, when phenomena accessible to external observation, such as changes in appearance and behavior, are distinguished and designated as *symptoms*, then the phenomenal domain within which they are observable is the system of interaction and communication.

Thus, to emphasize this point from a systems-theoretical viewpoint once again, the phenomena denoted as *symptoms* are elements of both the communicative system and the psychic system. Corporeal phenomena and forms of behavior in and of themselves are never symptoms; only when observers communicate about them and identify them as "symptoms" do they become symptoms. Psychic and interactional phenomena that are not comprehensible within the framework of everyday communication stand a chance to be interpreted as symptoms. As observable signs they can then be read in such a way that the existence of states, events, or processes outside the social system or communication can be inferred from them. Symptoms are explained by a generative mechanism located in a different phenomenal domain, such as the body, which is not directly observable without technical aids. Here the expert, the healing adept, can assume the role of someone who confers meaning.

The Explanation of Illness

Regarding illness as a change within somatic space does not (and never did) necessarily imply localizing its cause—that is, its generative mechanism—within the body. The earliest sources from the Babylonian and Egyptian periods, and similarly those from Greek and Roman antiquity, present two competing models for understanding or explaining illness which today still rule the discussions and controversies surrounding the interdependence of organic, psychic, and social factors in the genesis of illness. Without a doubt most influentially in our century, Karl Jaspers made the distinction between "understanding" and "explaining" the central dimension of differential diagnosis in psychiatry in order to distinguish physiologically conditioned from psychologically conditioned disorders.[12]

So-called "magical-animistic" medicine directly connects the pathogenic effect of the demons to violations of social taboo. Ancient, theurgic models of disease genesis, and likewise medieval monastic[13] medicine with its iatrotheology, either viewed sickness as divine punishment for transgressions or perceived it teleologically as the path to salvation.

In models of this kind, illness is interpreted as a sign that refers to another, in this case noncorporeal, phenomenal domain, wherein the change or deviation explaining the appearance of symptoms (that is, the cause of or agency responsible for the symptoms) is located. In cases where sickness is considered the gods' punishment, it serves as a sign of a postulated transgression of social norms. This way things come full circle: observable yet incomprehensible phenomena are attributed to past, still unobservable though re-understandable forms of behavior by means of two imaginary or at least not directly observable links, namely *illness* and *divine penalty*. Thus the domain of what is comprehensible is expanded indirectly through unobservable partners of interaction: symptoms can be read with reference to their meaning and hence become communication.[14]

Gods and demons too, are but "human"; at least they follow the same rules of interaction and communication as do humans. Their actions are understandable, even though they are not directly observable. When observable phenomena (symptoms) are evaluated negatively and are experienced as punishment, divine intent can be construed. Illness is judged morally: the sick person is guilty, as evidenced by his illness.

The steps for understanding symptoms that the observer must perform in his distinctions and evaluations conform to the following pattern: an observable but incomprehensible phenomenon is first traced back to an unobservable but comprehensible event and eventually to an unobserved but theoretically observable and understandable behavior. The question posed by the presence of any incomprehensible phenomenon to those participating in communication—"Does this phenomenon have a communicative meaning?"—can thus be answered in the affirmative. It possesses a meaning on the social level through which communication becomes possible again.

Many etiological[15] concepts in science-based medicine appear to follow a similar logic. They, too, are patterned after social rules and their inherent victimizer/victim distinction. Yet the blame (cause) is sought not within the social system but in the domain of the biological system. When an infection is considered the cause of a disease, the *disease-causing agent* is the wrongdoer, which then—following the model's logic—must be sought, found, and apprehended.[16]

In general terms, scientific explanatory models can only describe generative mechanisms observable by scientific methods. We have come full circle: Scientific concepts of illness can only attribute physical, chemical, and biological phenomena to physical, chemical, and biological phenomena. Considered by themselves, they consequently are free of meaning within a communicative context. Hence this form of biological explanation leaves all social and psychic phenomena unmarked. Therefore, on the social level, biological explanations in general lead to the conclusion that symptoms are not interpreted as communication, so the patient is absolved of blame. Instead the (objective) "cause" is invented or found; that is to say, a symptom-generating mechanism is constructed outside the communicative system.

An observable, incomprehensible phenomenon is traced back to an unobservable yet explainable event or process. In this way an explanation is constructed from which communication can proceed. The question of whether the initially incomprehensible symptom has communicative meaning is negated.

All psychosocial and psychosomatic disease models confront the problem that they aim to interconnect events in different phenomenal domains (psychological, social, and biological) in terms of causal, deterministic if-then relations.

Viewed through the theory of operationally closed systems, diseases are structurally determined reactions of the somatic system to disturbances (perturbations) triggered by changes in the environment. In this respect, illness is not different from health; in the case of health, too, the biological system (body) reacts to changes in the environment according to its structures. Biological systems make the distinction between alive and not-alive; as a result they maintain the system/environment distinction. Either the body system interacting with its environment succeeds in continuing its operations and thus lives on, or it does not succeed and subsequently dies, and its structures connected to the performance of life-preserving processes dissolve.[17]

Distinguishing between *sick* and *nonsick* with respect to particular corporeal reactions and the construction of "disease entities"[18] hence is not determined biologically but socially. The distinction is a feature of observation (of the "map") but not of the phenomena observed (not of the "landscape"). Moreover, the localization of causes for those observed phenomena in a biological, psychic, or social system or in one of its environments is a socially determined "punctuation"[19] of the corresponding system / environment interaction.

'Health' as Unmarked Space, State, or Content

Applying Spencer-Brown's laws of form, one can notice that in therapy illness is for the most part the marked space, state, or content that is signified and assessed and for which explanations are constructed. From this marked space, state, or content a nonmarked (i.e., unmarked) space, state, or content is distinguished which is designated *health* but not defined in terms of content. "Unmarked space" means that there is no positive distinguishing feature; in other words, health can only be recognized when illness is absent, that is to say, when the distinctive features characteristic of *sickness* are not present.

This explains why no one (not even the World Health Organization) thus far has ultimately succeeded in establishing a satisfying and generally acceptable definition of health, though it seems to be relatively easy to agree on a definition of illness.

There are, nonetheless, a few exceptions where health is the marked space. In cases such as fractures and other disorders pertaining to mechanically effective parts of the body, it is possible to specify the meaning of *health*: It is the original, intact state prior to injury. Only when physical functions can be assessed according to mechanical rules is it possible to define health in positive terms.

Therapist and Patient as Observers

The different vantage points for observing, understanding, and explaining that the patient and therapist have at their disposal or that are attributed to them have always determined the complementariness of their relationship.

The patient, and likewise the therapist, can directly observe symptoms in the domain of interaction; this is a phenomenal domain open to either view. When symptoms are psychic events, however, direct observation is possible for one observer only—the patient, who, for example, suffers from pain.

On the opposite side, the therapist has the possibility of looking at the patient from an external viewpoint. The therapist's diagnostic gaze permits him to distinguish phenomena the patient himself may not either perceive or be able to understand or explain meaningfully: What, for instance, does a pressing pain in the right side of the lower abdomen indicate? In the extreme case the therapist of today can penetrate boundaries impenetrable to the patient; he "x-rays" him. The therapist sees what the patient cannot.

Moreover, the therapist's special training also provides him with explanatory knowledge of each disease accepted by the healing profession, be it knowledge of the generative mechanisms that contemporary, scientifically oriented medicine constructs in the corporeal domain or an understanding of the strength and weaknesses of the gods, which, in theurgic or "magical-animistic" medicine, the therapist, as the patient's defender, can utilize in the divine "trial."

The therapeutic-technical question "How do I fight spirochetes most effectively?" differs little in its inherent logic from the question "How do I appease the gods?" To answer either question, the therapist is expected to be familiar with the mechanisms that are considered the cause of or culprit in the occurrence of symptoms and located in a domain of the world that, though unobservable, is regarded as real. And since he is able to interpret signs of developing symptoms that are for others meaningless, he probably also has communicative access to this unobservable domain, allowing him to intervene purposefully and beneficially.

This exclusive ability to see, explain, and understand what the patient cannot see, explain, and understand differentiates the therapist as an observer (likewise the priest, who also promises salvation) from the patient as an observer (seeking cure). For the sake of preserving this expert status, the profession was for centuries organized like a sect and its knowledge taught as a secret lore; even today it is for the most part handled this way. The similarity between the roles of therapist and priest, which for many centuries were not at all separated, also results from this supposedly different observational viewpoint: from the Black Plague to AIDS, times of major disease have always been boom times for preachers of repentance.

But the patient, too, possesses, as an observer (of himself), an exclusive status: he is able to make statements that cannot be falsified by other observers. Only he has access to what he is experiencing. This leads to far-reaching objectivity problems as to whether a symptom is or is not indeed present (cues: malingerer, hypochondriac, "retirement neurotic").

Behavioral and Communicative Modes as Symptoms

More than somatic phenomena, behavioral and communicative modes are scrutinized for meaningfulness according to the guiding distinction between comprehensible and incomprehensible. Those not adhering to certain grammatical rules in speech or not using the established vocabulary cannot take part in communication. But following formal rules of communication is not enough; content rules must be followed as well: anyone who attributes meaning in a manner running counter to the current conception of reality is labeled *losing his senses* or *losing his mind* (i.e., mad). Those who behave incomprehensibly in direct interaction drop out of communication. Such utterances are not connectable since they rest on descriptions, evaluations, and explanations not shared by others.

That therapists are assigned the responsibility of healing such "mad" people ensues logically from the aforementioned distinction that identifies symptoms as phenomena not immediately comprehensible. As early as ancient Greek medicine, madness was interpreted accordingly as a sign of illness, and the Occidental art of healing was occupied with the treatment of people who behaved conspicuously and strangely.[20] Since the conception of illness at that time was still very closely bound to the idea that sickness is divine punishment for moral transgressions, it was almost a natural outcome that a specialized art of healing competent in communicating with divine authorities would emerge.

In the fifteenth century, the first hospitals exclusively dedicated to the insane opened in Spain. In the rest of Europe, insane people served as entertainment for an educated audience, yet they were left free. Only after 1650 were they made outsiders and incarcerated all over Europe.

Where comprehensibility provides the guiding distinction, the criteria of that distinction are naturally quite weak: what is comprehensible for one observer remains incomprehensible for another. Much information on the actual social, cultural, economic, and historical context in which a particular behavior occurs is needed to understand it (that is, to comprehend the internal perspective of the communication participant). And understanding behavior is for the most part connected to its valuation. "Common sense" is the yardstick for comprehensibility. If comprehensibility is defined in reference not to form but to content, it fulfills a function on the level of interaction which corresponds to the function of ethical or moral values: Behaving or not behaving in a certain way is ensured by prescriptive rules. Any deviant behavior can be assessed as a symptom. Even "sinning" can be incomprehensible. For the distinction between "mad" and "bad" there are no unequivocal distinguishing marks. Both types of deviant behavior increase complexity on the social level.

Thus, one was confined to the European asylums not for therapeutic purposes but because one could not, or was not allowed to, participate in everyday social life. In the asylums one would encounter a colorful variety of highly dissimilar individuals: "the poor and disabled, the elderly poor, beggars, the work-shy, those with venereal diseases, libertines of all kinds, people whom families or the royal power wished to spare public punishment, spendthrift fathers, defrocked priests; in short, all those who, in relation to the order of reason, morality, and society, showed signs of 'derangement.'"[21] The guiding distinction for being admitted to one of those "hospitals" was sinful/not-sinful. And in the just-emerging bourgeois society, the one cardinal sin was "sloth."[22]

With the French Revolution these asylums were opened for all except the insane. For them, incarceration practices were intensified; the insane were subjected to moral and social control; and recovery was supposed to be achieved by restoring a sense of guilt and gratitude.[23]

It was (and still is) impossible or at least difficult to discriminate between therapy, punishment, and education. Moreover, assuming that usage determines the meaning of a term, it makes sense in the logic of the social-control function of therapy that in fifteenth-century German dialects *heilen* (to heal) as a loan-translation of *sanare* came to mean "to castrate." *Insanus* was the special term for animals that were too wild to put in front of a plow, and accordingly *sanare* meant "to take away the savageness from the male animal by cutting off its testicles."[24]

The development of the "clinical gaze," which enabled the description of correlations between intracorporeal processes and externally perceivable symptoms, ensured that those patients who could be diagnosed with a physiological finding were absolved of guilt and moral condemnation. The clinical gaze served as a sufficient source for explaining symptoms. The search for an explanation within the communication system could be halted at this point.

Regarding certain behaviors as symptoms of disease implies that someone behaving in such a way cannot be held responsible or blamed for the behavior. Illness is thus not viewed as a result of decisions made by an autonomous subject or of punishment or reward. When "objective" (physiological) explanations are lacking, this kind of absolution cannot be bestowed.

According to the *responsible/not-responsible* distinction, different social function-systems are put in charge of treating people who exhibit deviant behavior. Anyone violating prescriptive laws either is "responsible," in which case he is subjected to educational, disciplinary, and legal measures, or displays symptoms, in which case he is "not responsible" and becomes the object of care, nursing, and therapy. The psychiatric expert, with his distinction *can be blamed / cannot be blamed*, is all too often given the task of declaring either the courts or the clinics the appropriate institution for the subject.

Whenever illness is assumed as an explanatory, that is, generative mechanism, the observable behavior is causally explained by unobservable processes. Or again, from the viewpoint of systems theory, phenomena belonging to the domain of interaction and communication are not interpreted as meaningful within the realm of interaction but are traced back to changes in the environment of the social system. In this way a context is created for direct interaction with the one who exhibits such behavior. In this framework the individual cannot be held accountable for his deeds since he is defined not as the *agent* but as the *victim* (of a disease).

With respect to the individual, the consequences of this difference can be very important. On the social level, however, the distinction between therapy and punishment is suspended at the point when the violation of social rules leads to decreasing predictability in interaction. When trust is no longer an effective mechanism in reducing complexity, mechanisms of control are introduced. Excluding *mentally ill people* who violate the rules of "direct interaction"[25] has the same controlling, complexity-reducing effect as incarcerating criminals. The official rationale and objective for the exclusion are all that differ between the two cases (*therapy* and *punishment*).

The Invention of the Unconscious

Whereas the biological model of *mental illness* attributes the observable symptom (certain behavior) to an assumed, unobservable *illness of the brain* ("mental diseases are brain diseases"),[26] the psychological model ascribes mental illness to an assumed, unobservable *psychic illness*. In the first case one searches the organismic system for structural or functional abnormalities that explain the development of symptoms; in the second case the psychic system is examined.

Once the scientific "fact"[27] that progressive paralysis is caused by a spirochetal infection was successfully established, biological psychiatry experienced a major boom. And quite in line with the expansion of the clinical gaze into the region beyond the boundary of the skin, most scientific energy (at least, most of the money) within contemporary psychiatry seems to be invested in the search for biological functional and structural changes.

In the last hundred years, however, an alternative (second) way of searching for internal causes (i.e., generative mechanisms) has evolved with psychology and psychopathology. Here the generative mechanism for symptoms is sought in the psychic system; accordingly, the assumed *illness* is a *psychic illness* (thus not a cerebral illness).

Let us consider the psychoanalytical model as an example of a psychological conception of illness.

Developed in the context of neurological praxis, this conception displays the typical observational and distinguishing schemata described above. In communicating with his patients, Freud interpreted their incomprehensible symptoms by tracing them back to a mechanism unobservable by him or the patients, that is, to an unconscious mechanism. He divided, if you will, the realm of the psyche into that which is accessible to self-observation and that which is not. He considered the unconscious an intelligible, autonomous partner for interaction, with its own motives and goals. The generative mechanism he constructed emulates the hermeneutic model, that is, one of searching for and constructing meaning.

In contrast to the biological disease models that attribute incomprehensible behavior (symptoms) to causal links situated outside the social system, Freud's theory turned unintelligible behaviors into intelligible ones by expanding the realm of comprehensibility. Freud thereby performs a step analogous to those in theurgic and "magical-animistic" models: they, too, extended the domain of intelligibility by constructing imaginary, understandable partners of interaction. The motives of the gods and demons are genealogically linked to the motives of the unconscious—an imaginary, seemingly almighty partner of interaction that satisfies wishes when they are deliberately prohibited and forbids their gratification when consciousness deems them permissible.

By contrast, disease models based on theories of learning seek, like biological ones, to explain the development of symptoms. They operate with psychic, instead of biological, generative mechanisms. The models of *mental illness* and *psychic illness* have in common the fact that they locate the cause of unintelligible behavior in changes outside the communicative system.

A Systems-Theoretical Model of Symptom Development

Explaining incomprehensible behavior by reference to structural or functional changes in the brain or in the psyche constitutes a form of mythological explanation:[28] the constructed generative mechanisms are related not to observable phenomena but to the imaginary structures and processes of an unobservable domain.

This recourse to mythological explanations can be accounted for by the fact that people's behavior and speech are usually interpreted in communicative settings with respect to the message conveyed. Now if a certain behavior appears nonsensical and meaningless, then it cannot be read on the basis of existing rules of communication. One draws the only (apparently logical) conclusion that its generative mechanism must be sought outside the communication system, that is, in its environment (biological or psychic system).

Making systems theory and communication theory the basis of our model leads first to the question of how one actually accounts for understanding. To explain understanding, we need a generative mechanism; understanding cannot simply be taken for granted and left unscrutinized. With respect to incomprehensible behavior, such a perspective leads to the question of whether the termination of communication might be explained as a consequence of communication.

The appeal of this approach is that it explains events and processes occurring within the systems of interaction and communication by reference to events and processes within those systems. This matches the parameters of a theory of operationally closed autonomous systems that are structurally defined and are not determined behaviorally by changes in the environment.

Behavioral and communicative symptoms can be explained in this model as outcomes of communication and interaction; they remain, nonetheless, incomprehensible.[29]

Such a theoretical model is the foundation of systemic therapy, whose techniques were developed primarily in the treatment of families with psychotic symptoms.[30] The system to be treated is now the social system, and the development of symptoms is explained by its rules. The psyche and body of each family member, and the internal dynamics between the members, are environments of the symptom-producing system; they can limit but not determine the developmental possibilities of the social system, as is assumed in the "mythological" explanations presented earlier.

Interventions in Self-Referential Systems

The task of therapy arises from the distinction between *sick* and *nonsick*. Therapy is supposed to engender and promote the crossing of the boundary from what is termed the *sick* state to the *healthy* state.

Conceptions about how such a therapeutic effect can be achieved are for the most part patterned after the underlying pathogenic explanations. If health is taken for granted by the observer, he does not construct a generative mechanism to account for its genesis. *Healing* is therefore largely conceptualized as eliminating pathogenic causes but not as creating conditions for health.

When the development of symptoms is explained as an intrusion of "pathogens" into the body, exorcising those "malefactors" (be they demons, parasites, cocci, viruses, or spirochetes) becomes the conclusive conception of therapy. Exorcistic rites and cathartic psychotherapeutic methods follow the same principle of purification as do antibiotic treatments for bacterial infections, the removal of inflamed appendixes, and the lancing of abscesses.

When imbalance or disturbed harmony is considered the main characteristic of illness, therapy focuses on restoring this harmony. Hippocratic medicine and subsequently Galen's medicine sought, for instance, to restore the proportions of bodily fluids to a harmonic equilibrium (syncracy, eucracy) by dietary means. That is to say, the treatment's underlying principle was that of counteractive measures (*contraria contrariis*).

When the explanation of illness is located on a hermeneutic level, therapy aims to change the illness-causing conditions of interaction (either with gods or with humans). When taboo violations have led to social exclusion, not only must the raging demon be appeased or exorcised, but social reintegration rituals must be performed as well. If illness is punishment, then absolution is needed; the pilgrimage to Lourdes promises salvation on both the physical and psychic levels.

Psychoanalysis that regards unconscious conflict as the generative mechanism of psychic illness seeks to let the patient become aware of the unobservable part of his psychic life so that the unconscious conflict will lose its pathogenic efficacy. With regard to understanding, the difference between therapist and patient disappears.

Therapy looks somewhat different when we seek to conceptualize it in terms of recent systems theory as an intervention into self-referential, autopoietic systems (organism, psyche, social system).

Operationally closed systems are structurally determined, meaning that an instructive interaction between a therapist and a system designated as *sick* is not possible.[31]

The inside/outside distinctions by which most therapeutic concepts are guided, however, are properties not of an observing system but of its description. For example, the body does not distinguish between inside and outside (as an external observer could); the body's different operations are determined by biological structures, and they simply occur the way they do. For biological systems, there is no environment—no distinction between self and non-self; for that an external observational perspective would be required.

The distinction made by the operations of the biological system is not between self and non-self but between—as Francisco Varela stresses—"self and non-sense." Either the organism does have an operational mode for reacting to a particular event, in which case it will react to such an event—the organism operates as if the event makes "sense"—or else, if there is no preexisting operational pattern, no reaction will ensue—the event remains "nonsensical" to the organism.[32]

"Meaning" (or "sense") in this context signifies that a system responds to an event with an internally defined reaction. Corporeal processes designated by an outside observer as *illness* are thus reactions determined by the biological structure of the body to events that are either "meaningful" or "without meaning" to the organism. Therapeutic possibilities are restricted to the use of those physiological, structurally determined modes of operation.

From an external observer's point of view, two types of pathogenesis can be distinguished in principle. In one case, the body interprets as "meaningless" events that, for maintaining a state of health, should in fact be read as "meaningful," and then operations and processes do not take place which could prevent the development of symptoms. In the other case, events are interpreted as "meaningful" when, for the benefit of the body, they should be taken as "meaningless," and then operations and processes occur that produce symptoms.

Therapeutic interventions into the organism as a self-referential system can aim to produce or eliminate events that are "meaningful" to the body, or they can see to it that recurring events lose or gain "sense" for the body. Thus by using the organically determined procedural patterns, the necessary somatic reactions for eliminating illness can be triggered, or those reactions that cause illness can be prevented.

The same principle applies to interventions on the level of psychic or social systems. Problems are generated either because necessary operations are not carried out or because unnecessary ones are. Therapeutic or counseling intervention can be effective only on the level of modifying the attribution of meaning (here we use the term less metaphorically), that is, on the level of changing individual or collective systems of meaning. By means of communication, therapy can achieve two positive outcomes: the prevention of problem-generating operational modes on the psychic or social levels, and the promotion of operations on the two levels necessary for the maintenance of a problem-free state.

Since systems of meaning arise out of communication, they can also be questioned by communication.

Finite Therapy—Systemic Therapy

When ways of behaving and communicating are seen as symptoms, the shortest way to reach the therapeutic objective is a mode of interaction and communication between therapist and patient in which the initial assumptions about the patient (descriptions, evaluations, and explanations) lead to either of two outcomes: either to conclusions that no longer confirm but rather falsify the premises or, on the level of interaction, to behavior that no longer leads to repetitive cycles, to the same old patterns of interaction.

In either case, self-fulfilling loops turn into self-negating loops; that is, there is an opening up of the operational closure by which a symptom is produced and maintained as an *eigen*-value or *eigen*-structure.[33] Hence the symptom is no longer produced.

Defining the therapeutic goal as an unmarked state (that is, as the absence of symptoms, problems, suffering) turns out to be far more useful to the therapist than following the normatively predetermined aim of striving toward *health* as a marked state. Since there is no objective criterion for discerning *health*, therapy based on *health* risks becoming interminable—a problem of psychoanalysis, for *real genital maturity* is quite difficult to ascertain.

For the patient, the situation looks different: the therapeutic goal must become the marked state. Only if he defines a distinguishing mark by which he recognizes that he is doing well or by which he can call himself *healthy* or *free of suffering* can he make the transition from the state of illness to the healthy or nonsuffering state.

When a patient visits a therapist for the first time, he usually reports certain symptoms. The therapist's activity focuses on answering the question of what the signs (words, concepts, etc.) used by the patient signify on the behavioral level. That is to say: What behaviors must the patient exhibit, and in which social and situational contexts, so as to find himself in a state that he then can describe in words he now uses to signify his symptoms? What partners of interaction are involved? What behaviors does the patient describe, and how does he evaluate and explain them? How does he subsequently behave given his description of himself and the interaction? What are the differences on the level of behavior and interaction, respectively, if the outcome of interaction is a state labeled *nonsuffering*? Or in other words: What would the patient and others do differently, so that he could call himself *healthy* or something similar?

Thus the therapeutic dialogue is about jointly developing ideas regarding the patient's forms of distinctions and significations, that is, ideas about his part in producing and maintaining symptoms. How does he—and anyone participating in the interaction—manage to prevent change, that is, the way in which they wind up in a self-fulfilling loop?

Therapy then means interrupting this self-fulfilling loop, transforming the tautology into a paradox: the self-fulfilling loop becomes a strange loop.[34]

It goes without saying that such an approach is much more effective when not just one individual but all who prolong a certain pattern of interaction (game) participate in therapy. If one or several of them behave differently, the pattern of interaction that has formed around the symptom changes, so the symptom itself changes as well.

Reference Matter

NOTES

Baecker, Introduction

1. Social-systems theory's turn to distinction theory was heralded in Niklas Luhmann's "Frauen, Männer und George Spencer Brown," *Zeitschrift für Soziologie* 17 (1988): 47–71.
2. George Spencer-Brown, *Laws of Form*, first published in London in 1969; first American edition, New York, 1972; various reprints and paperback editions, all out of print; last limited American edition, Portland, Oregon, 1994; new German edition with new material, Lübeck, Germany, 1997. Only since the 1979 reprint paperback edition has Spencer-Brown added the hyphen to his name, presumably in order to tell librarians not to hide his book under "Brown, G. Spencer."
3. For discussion of the mathematical dimension of Spencer-Brown's theory, see the following works by Louis H. Kauffman: "Self-Reference and Recursive Forms," *Journal of Social and Biological Structure* 10 (1987): 53–72; "Ways of the Game—Play and Position Play," *Cybernetics and Human Knowing* 2, no. 3 (1994): 17–34; "Arithmetic in the Form," *Cybernetics and Systems* 26 (1995): 1–57.
4. See Heinz von Foerster's review of *The Laws of Form* in *Whole Earth Catalogue* (Spring 1969), p. 14; Francisco J. Varela, "A Calculus for Self-Reference," *International Journal of General Systems* 2 (1975): 5–24; and Francisco J. Varela and Joseph A. Goguen, "The Arithmetic of Closure," *Journal of Cybernetics* 8 (1978): 291–324.
5. Spencer Brown, *Laws of Form*, p. 3. I quote the 1972 edition.
6. Ibid., p. 1.
7. Ibid., pp. 1 and 2.
8. See ibid., chaps. 11 and 12, on "equations of the second degree" and "reentry into the form."
9. Ibid., p. 65.
10. See Niklas Luhmann, *Die Gesellschaft der Gesellschaft* (Frankfurt a.M., 1997).
11. See, e.g., Francisco J. Varela, Antonio Coutinho, Bruno Dupire, and Nel-

son N. Vaz, "Cognitive Networks: Immune, Neural, and Otherwise," in Alan S. Perelson, ed., *Theoretical Immunology*, vol. 2 (Redwood City, Calif., 1988), pp. 359–75.

12. Yet note the earlier discovery of the laws of form by Buddhist practice and Indian wisdom. One may recall Don Juan making Carlos Castañeda "cross" his all-too-self-evident distinctions, or the famous figure of Old Coyote doing the same in Indian tellings. See, e.g., the stories retold in Barry Lopez, *Giving Birth to Thunder, Sleeping with His Daughter: "Coyote Builds America"* (New York, 1977). The knowledge of the "laws of form" in traditional society is labeled, and thereby handed down, as "mystical knowledge." For the self-conditioning of social systems, see examples in Niklas Luhmann, "'Distinctions directrices': Über Codierung von Semantiken und Systemen," in his *Soziologische Aufklärung 4: Beiträge zur funktionalen Differenzierung der Gesellschaft* (Opladen, 1987), pp. 13–31; and of course in Luhmann, *Die Gesellschaft der Gesellschaft*.

13. See Auguste Comte, "Plan des travaux scientifiques nécessaires pour réorganiser la société," in *Système de politique positive* (Paris, 1854), pp. 47–176.

14. For this list of five "unit-ideas" of sociology, see Robert A. Nisbet, *The Sociological Tradition* (New York, 1966).

15. For the corresponding epistemology, see the two papers "Notes on an Epistemology for Living Things" and "Objects: Tokens for (Eigen-)Behaviors," in Heinz von Foerster, *Observing Systems* (Seaside, Calif., 1981), pp. 258–71 and 273–85.

16. Anthony Giddens's concept of "structuration," Pierre Bourdieu's concept of "field," Harrison C. White's concept of "control," and Niklas Luhmann's concept of "system" all describe processes engendering their own reproduction. See Anthony Giddens, *The Constitution of Society: Outline of the Theory of Structuration* (Berkeley, Calif., 1984); Pierre Bourdieu, *Distinction: A Social Critique of the Judgement of Taste*, trans. Richard Nice (Cambridge, Mass., 1984); Harrison C. White, *Identity and Control: A Structural Theory of Action* (Princeton, N.J., 1992); and Niklas Luhmann, *Social Systems*, trans. John Bednarz, Jr., with Dirk Baecker (Stanford, Calif., 1995).

17. See, e.g., Hans Ulrich Gumbrecht, "Form Without Matter vs. Form as Event," *Modern Language Notes* 111 (1996): 578–92.

18. With respect to paradoxes of intension and extension or of the elementary and the universal, see Ranulph Glanville, "Beyond the Boundaries," in R. F. Ericson, ed., *Improving the Human Condition: Quality and Stability in Social Systems* (Louisville, Ky., 1979), pp. 70–74; and Ranulph Glanville and Francisco J. Varela, "Your Inside Is Out and Your Outside Is In," in G. E. Lasker, ed., *International Congress on Applied Systems Research and Cybernetics*, vol. 6 (New York, 1981), pp. 638–41.

19. This is not the place to discuss the possibility of "reading" deconstruction via Spencer-Brown's calculus. Yet the similarities between deconstruction's and Spencer-Brown's (and systems theory's, for that matter) interest in the relationship

between paradox and distinction are striking. The deconstructive reading of texts often reads like an exposure of unmarked states shaded by indications and of operations of distinctions drawn by observers hidden behind the veils designed to make them invisible. Spencer-Brown's calculus may prove to be the "method" from which deconstruction shies away because of its self-imposed disbelief in conceptual self-exposure. Interestingly, though, this disbelief is not recommended when one deals with communication. See Jacques Derrida, "Postface: Vers une éthique de la discussion," in *Limited Inc.*, ed. and trans. Elisabeth Weber (Paris, 1990), pp. 199–285.

20. This was first spelled out for biology in Humberto R. Maturana and Francisco J. Varela, *Autopoiesis and Cognition: The Realization of the Living* (Dordrecht, 1980), but was anticipated by the idea of "recursivity" developed in second-order cybernetics, the cybernetics of observing systems. See Heinz von Foerster, "Cybernetics," in Stuart C. Shapiro et al., eds., *Encyclopedia of Artificial Intelligence*, vol. 1 (New York, 1987), pp. 225–27.

21. The distinction between symbolic and formal analysis was once introduced by structural analysis. See, for instance, Michel Serres, "Structure et importation: Des mathématiques aux mythes," in *Hermès I: La communication* (Paris, 1968), pp. 21–35. Structural analysis was already in search of the mechanisms producing the phenomena it wanted to explain. Yet it remained faithful to some remnant of symbolic analysis in that it accepted a distinction between "deep structures" and "surface structures." The distinction theory of form replaces this distinction by the distinction between a paradox and its "enfoldment."

22. Heinz von Foerster therefore proposes a new version of the "metaphysical postulate": "Only *those* questions which are in principle undecidable, *we* can decide." See his "Through the Eyes of the Other," in Frederick Steier, ed., *Research and Reflexivity* (London, 1991), pp. 63–75.

23. For a corresponding notion of complexity, see Niklas Luhmann, "Haltlose Komplexität," in his *Soziologische Aufklärung 5: Konstruktivistische Perspektiven* (Opladen, 1990), pp. 59–76.

24. Bertolt Brecht's phrase "Wer wen?" (Who whom?) then echoes all fundamentalist, including totalitarian, societies.

25. Since the publication of this volume in German in 1993, further steps have been undertaken: with respect to law and art, by Niklas Luhmann in his works *Das Recht der Gesellschaft* (Frankfurt a.M., 1993) and *Die Kunst der Gesellschaft* (Frankfurt a.M., 1995); and with respect to the form of the firm, by Dirk Baecker in his *Die Form des Unternehmens* (Frankfurt a.M., 1993).

26. See the idea of *opposition hiérarchique* in Louis Dumont, *Essais sur l'individualisme: Une perspective anthropologique sur l'idéologie moderne* (Paris, 1983); and the hierarchical reconstruction of Derrida's *supplément* and Girard's mimesis, in Jean-Pierre Dupuy and Francisco J. Varela, "Understanding Origins: An Introduction," in *Understanding Origins: Contemporary Views on the Origin of Life, Mind and Society* (Dordrecht, 1992), pp. 1–25.

27. That the observer hides inside the unmarked state is sometimes assumed in linguistics. See Linda R. Waugh, "Marked and Unmarked: A Choice Between Unequals in Semiotic Structure," *Semiotica* 38 (1982): 299–318.

28. The suspicion remains that Hegel's dialectics uses the "principle of non-identity" (see, e.g., Michael Kosok, "The Formalization of Hegel's Dialectical Logic: Its Formal Structure, Logical Interpretation and Intuitive Foundation," *International Philosophical Quarterly* 6 [1966]: 596–631) only in order eventually to strengthen the return to unity.

29. Bateson once assumed that Spencer-Brown's "cross" is the equivalent of the operator of negation. "No, it ain't," was Spencer-Brown's answer when he learned of this assumption. The story of this incident, which happened at the G. Spencer-Brown-Aum-Conference at the Esalen Institute, California, in 1972, is retold by Bradford P. Keeney, *Aesthetics of Change* (New York, 1983), p. 19 n. 7. Negations are second-order distinctions. They manage to indicate the unmarked state without turning it into the marked state, or, better, they manage to indicate the unmarked state by turning it into a marker rather than a cross, as Freud long ago observed in his terms. See Sigmund Freud, "Die Verneinung," in his *Das Ich und das Es: Metapsychologische Schriften* (Frankfurt a.M., 1992), pp. 319–25. See also Dirk Baecker, "Was leistet die Negation?" in Friedrich Balke and Joseph Vogl, eds., *Gilles Deleuze—Fluchtlinien der Philosophie* (Munich, 1996), pp. 93–102.

30. The most important candidate being, of course, Gotthard Günther's essays on the foundation of an operational dialectics. See, e.g., Gotthard Günther, "Cognition and Volition: A Contribution to a Cybernetic Theory of Subjectivity," in his *Beiträge zur Grundlegung einer operationsfähigen Dialektik*, vol. 2 (Hamburg, 1979), pp. 203–40. See also Elena Esposito's and Rudolf Kaehr's recommendations to proceed from Spencer-Brown to Günther, in Dirk Baecker, ed., *Kalkül der Form* (Frankfurt a.M., 1993), pp. 96–111 and 152–96.

31. See, e.g., Spencer-Brown's concept of "unwritten crosses," emphasized by Matthias Varga von Kibéd, "Wittgenstein und Spencer Brown," in Paul Weingartner and Gerhard Schurz, eds., *Philosophie der Naturwissenschaften: Akten des 13. Internationalen Wittgenstein Symposiums* (Vienna, 1989), pp. 402–6. Social-systems theory's communication analysis may itself be understood as an attempt to write the hitherto "unwritten cross" of social phenomena.

32. See Spencer Brown, *Laws of Form*, pp. 58–62.

33. See Niklas Luhmann, "Zeit und Gedächtnis," *Soziale Systeme: Zeitschrift für soziologische Theorie* 2 (1996): 307–30. This indeed is one important part of the architecture of Luhmann's theory of society, now presented in his *Die Gesellschaft der Gesellschaft*.

34. Keeping in mind that the distinctions between the modern and the traditional, between the European and the non-European, and between different disciplines of science are all very peculiar, providing for forms that only recently began to be observed with respect to their paradox, contingency, and continence.

35. With respect to distinction and asymmetry, see Robert Hertz, "La prééminence de la main droite: Étude sur la polarité religieuse," *Revue Philosophique* 68 (1909): 553–80.

36. For similar arguments, see Harrison C. White, *Identity and Control: A Structural Theory of Action* (Princeton, N.J., 1992), especially his "calculus of boundaries," pp. 127–28; and his paper "Interfaces," *Connections* 5 (1982): 11–20.

Luhmann, The Paradox of Form

1. Regarding *eidos/morphē* and the phenomenology of intentional meaning, see especially Derrida's "Form and Meaning: A Note on the Phenomenology of Language"; with respect to *ousia*, see in particular "*Ousia* and *Gramme*: Note on a Note from *Being and Time*"; both are in Jacques Derrida, *Margins of Philosophy*, trans. Alan Bass (Chicago, 1982).

2. According to Josef Simon's *Philosophy of the Sign*, trans. George Heffernan (Albany, N.Y., 1995), every solution sought in the immediacy of understanding or in pragmatic constraints is only ever tentatively acceptable and can at any time be overridden by the need for further signs.

3. George Spencer-Brown, *Laws of Form* (reprint, New York, 1979), p. 1. All page numbers in the main body of this essay refer to this edition.

4. I am restating here the following passage: "a distinction is drawn by arranging a boundary with separate sides so that a point on one side cannot reach the other side without crossing the boundary" (p. 1).

5. Without further clarification, Spencer-Brown even speaks of "motives." We are obviously in the midst of an already created world.

6. Louis Kauffman explicitly cites Spencer-Brown's introductory statement as an example for the form of re-entry. See Louis H. Kauffman, "Self-Reference and Recursive Forms," *Journal of Social and Biological Structure* 10 (1987): 53–72, see p. 58.

7. Ibid., p. 53. For a solution by means of self-indication, see also Francisco J. Varela, "A Calculus for Self-Reference," *International Journal of General Systems* 2 (1975): 5–24.

8. "Call the form of the first distinction the form. Let there be a form distinct from the form. Let the mark of the distinction be copied out of the form into such another form." Spencer-Brown, *Laws of Form*, p. 4.

9. Chapter 12 ("Re-entry into the Form") above all points to such an extended interpretation.

10. For a more detailed discussion of this point, see Niklas Luhmann, "Gleichzeitigkeit und Synchronisation," in his *Soziologische Aufklärung 5: Konstruktivistische Perspektiven* (Opladen, 1990), pp. 95–130.

11. The related formulation, "the narrator both in and not in the narration," is actually quite a familiar theme in literature and literary historiography. See

Dietrich Schwanitz, "Zeit und Geschichte im Roman—Interaktion und Gesellschaft im Drama: Zur wechselseitigen Erhellung von Systemtheorie und Literatur," in Dirk Baecker et al., eds., *Theorie als Passion: Niklas Luhmann zum 60. Geburtstag* (Frankfurt a.M., 1987), pp. 181-213; Dietrich Schwanitz, *Systemtheorie und Literatur: Ein neues Paradigma* (Opladen, 1990); and David Roberts, "Self-Reference in Literature," in this volume.

12. By the way, this is an observer who, luckily or unluckily for him, falls in love—as James Keys (alias George Spencer-Brown) reports—and thereby steps over into a different play of distinctions and different forms of dissolving the paradox of form. See James Keys, *Only Two Can Play This Game* (Cambridge, Eng., 1971).

13. In this regard see especially Heinz von Foerster, *Observing Systems* (Seaside, Calif., 1981). Also see the *festschrift* for Heinz von Foerster: Paul Watzlawick and Peter Krieg, eds., *Das Auge des Betrachters: Beiträge zum Konstruktivismus* (Munich, 1991); and Humberto Maturana, "The Biological Foundations of Self-Consciousness and the Physical Domain of Existence," in Niklas Luhmann et al., eds., *Beobachter: Konvergenz der Erkenntnistheorien?* (Munich, 1990), pp. 47-117.

14. According to Maturana, "Biological Foundations of Self-Consciousness."

15. This is also Kauffman's premise, and he reaches the following conclusion: "Therefore, self-reference and the idea of distinction are inseparable (hence conceptually identical)." See "Self-Reference and Recursive Forms," p. 53.

16. In the first volume of *Vorlesungen über schöne Literatur und Kunst* (Lectures on Literature and Art) Schlegel states: "But if one views all of nature as a self-conscious being, what would one think of the unreasonable demand put to her: to study herself by means of experimental physics?" August Wilhelm Schlegel, *Die Kunstlehre* (Stuttgart, 1963), p. 49.

17. See Peter J. Awn, *Satan's Tragedy and Redemption: Iblis in Sufi Psychology* (Leiden, 1983). Here the dissolution of the paradox lies in a new paradox: that the most loved angel experiences his damnation as an act of love that honors him above all others, and therefore he cannot repent of it.

18. Keys, *Only Two Can Play This Game*, p. 24.

19. The Greek term *diairesis* (from *haireo*, to grasp, to seize) does not permit one to distinguish between distinguishing and dividing.

20. Plato, *Sophist* 253 D.

21. For the transformation of an aristocratic *ragione signorile* into a specifically juridical conceptional technique with corresponding systematizations, see Aldo Schiavone, *Nascita della giurisprudenza: Cultura aristocratica e pensiero giuridico nella Roma tardo-repubblicana* (Bari, 1976).

22. See Rosalie L. Colie, *Paradoxia Epidemica: The Renaissance Tradition of Paradox* (Princeton, N.J., 1966); and Ulrich Schulz-Buschhaus, "Vom Lob der Pest und vom Lob der Perfidie: Burleske und politische Paradoxographie in der italienischen Renaissance-Literatur," in Hans Ulrich Gumbrecht and K. Ludwig

Pfeiffer, eds., *Paradoxien, Dissonanzen, Zusammenbrüche: Situationen offener Epistemologien* (Frankfurt a.M., 1991), pp. 259-73.

23. See A. E. Malloch, "The Techniques and Function of the Renaissance Paradox," *Studies in Philology* 53 (1956): 191-203; and Michael McCanles, "Paradox in Donne," *Studies in the Renaissance* 13 (1966): 266-87. For a contemporary reference, see, for example, John Donne's juvenilia *Paradoxes and Problems*, ed. Helen Peters (Oxford, 1980), as well as many of his later poems.

24. In the Introduction (section VI) of the *Critique of Judgment*, trans. Werner S. Pluhar (Indianapolis, 1987), Kant says: "It is true that we no longer feel any noticeable pleasure resulting from our being able to grasp nature and the unity in its division into genera and species that alone makes possible the empirical concepts by means of which we cognize nature in terms of its particular laws. But this pleasure was no doubt there at one time, and it is only because even the commonest experience would be impossible without it that we have gradually come to mix it in with mere cognition and no longer take any special notice of it" (p. 27). That is to say: the *genos* speculation is transformed back into a *genos* technique.

25. Kenneth J. Gergen, *Toward Transformation in Social Knowledge* (New York, 1982), p. 142.

26. This has been empirically validated! On the occasion of a lecture on moral code, which I opened with the paradoxical thesis that morality is not something good (since the difference between good and bad cannot be good itself), a philosopher present, referring to problems of racism, commented indignantly. Understanding the paradox, however, might have forced the question (distinction) of whether or not a moral assessment of this problem suffices.

27. They have been commended by, for example, Stephen Toulmin, *Cosmopolis: The Hidden Agenda of Modernity* (New York, 1990). The certainly most impressive work of postmodernity, by the way, originates also in this time period: Rabelais's *Gargantua and Pantagruel*.

28. For the corresponding controversy, see Ernest Nagel, *Logic Without Metaphysics* (Glencoe, Ill., 1956), pp. 247ff.; Ernest Nagel, *The Structure of Science* (New York, 1961), pp. 520ff.; Carl Hempel, "The Logic of Functional Analysis," in Llewellyn Gross, ed., *Symposium on Sociological Theory* (Evanston, Ill., 1959), pp. 271ff.; Gustav Bergmann, "Purpose, Function, Scientific Explanation," *Acta Sociologica* 5 (1962): 225-38.

29. "Son las paradojas monstruos de la veridad": Baltasar Gracián, in the "Discurso XXIII" of his treatise *Agudeza y arte de ingenio* (Huesca, 1649), quoted from the edition published in Madrid, 1969 (p. 224).

30. See Gilbert Chesterton, *Heretics* (London, 1905), p. 82, quoted from Hugh Kenner, *Paradox in Chesterton* (London, 1948), p. 14; see p. 25 for the paradox that difference necessarily presupposes sameness of what is different.

31. Thomas Browne, *Religio Medici* (1643), quoted from the edition of Everyman's Library (London, 1965), p. 10.

Roberts, Self-Reference in Literature

This chapter was previously published under the title "The Paradox of Form: Literature and Self-Reference," *Poetics* 21 (1992): 75–91.

1. Niklas Luhmann, "Weltkunst," in Niklas Luhmann, Frederick D. Bunsen, and Dirk Baecker, *Unbeobachtbare Welt: Über Kunst und Literatur* (Bielefeld, 1990), pp. 7–45; Niklas Luhmann, "European Rationality," in *Observations on Modernity*, trans. William Whobrey (Stanford, Calif., 1998), pp. 22–43; and George Spencer-Brown, *The Laws of Form* (New York, 1979). Page references in my text are to this 1979 edition of *Laws of Form*.

2. Niklas Luhmann, *Die Wissenschaft der Gesellschaft* (Frankfurt a.M., 1990), p. 526.

3. I understand the two aspects of the autonomy of art—social and world autonomy—as reflecting the complementary perspectives of Siegfried J. Schmidt, *Die soziale Organisation des Sozialsystems Literatur im achtzehnten Jahrhundert* (Frankfurt, 1989), and Dietrich Schwanitz, *Systemtheorie und Literatur: Ein neues Paradigma* (Opladen, 1990).

4. Luhmann, "Weltkunst," p. 42.

5. "Die Handlung der Freiheit, durch welche die Form zur Form der Form als ihres Gehaltes wird und in sich selbst zurückkehrt, heißt Reflexion." Johann Gottlieb Fichte, "Der Begriff der Wissenschaftslehre," quoted from Walter Benjamin, *Der Begriff der Kunstkritik in der deutschen Romantik* (Frankfurt, 1973), p. 16.

6. Philippe Lacoue-Labarthe and Jean-Luc Nancy, *The Literary Absolute: The Theory of Literature in German Romanticism* (Albany, N.Y., 1988), p. xxii.

7. Luhmann, "Weltkunst," pp. 11–12.

8. Gotthard Günther, quoted from Luhmann, "Weltkunst," p. 8.

9. Benjamin, *Der Begriff der Kunstkritik*, pp. 80–81.

10. Luhmann, "Weltkunst," pp. 42–45.

11. See Benjamin, *Der Begriff der Kunstkritik*, pp. 92ff.

12. Schwanitz, *Systemtheorie und Literatur*, pp. 111–12.

13. For fictionalization, see Schmidt, *Die soziale Organisation des Sozialsystems Literatur*, pp. 381–408, 427ff.

14. Schwanitz, *Systemtheorie und Literatur*, p. 154.

15. Bernhard Giesen, *Die Entdinglichung des Sozialen* (Frankfurt, 1991), p. 85.

16. Schwanitz, *Systemtheorie und Literatur*, p. 174.

17. Luhmann, "Weltkunst," p. 42.

18. See David Roberts, *Art and Enlightenment: Aesthetic Theory After Adorno* (Lincoln, Neb., 1991), pp. 163–73.

19. The Magritte quotation is taken from another master of the paradoxes of self-reference: Douglas R. Hofstadter, *Gödel, Escher, Bach: An Eternal Golden Braid* (Harmondsworth, Eng., 1980), p. 706.

Luhmann, Sign as For

1. For information n Galen and the medieval commentaries on Galen, see Per-Gunnar Ottoson, *Scholastic Medicine and Philosophy: A Study of Commentaries on Galen's Tegni (ca. 1300–1450)* (Naples, 1984), pp. 195ff. This concept of the sign is obviously supported by an epistemology that portrays the human mind (*mens*) as reacting passively to impressions such that whatever is given in perception can then be interpreted as a sign for something else.

2. For a continuation of this type of semiotic theory, expressed in the language of (ontological) metaphysics, see especially Josef Simon, *Philosophy of the Sign*, trans. George Heffernan (Albany, N.Y., 1995). According to Simon, however, it is no longer possible to make present what the sign signifies. Herein lies a sharp break with tradition. The traditional idea of the accessibility of the signified now is replaced either by that of interpreting signs through other signs or—in borderline cases that are always transitory—by that of immediate understanding. We will return to this point.

3. The explanation given for this distinction, namely *quae significant / quae significatur*, sounds quite modern. However, one has to take into account the explanation's presupposition of a world of meaningful things.

4. Plato, *Cratylus* 390 E.

5. See Bernhard Giesen, *Die Entdinglichung des Sozialen: Eine evolutionstheoretische Perspektive auf die Postmoderne* (Frankfurt a.M., 1991), especially p. 142.

6. Following jurisprudential usage, one could speak of the "category of illusory reference." See Julius Stone, *Legal Systems and Lawyers' Reasoning* (Stanford, Calif., 1964), pp. 235ff. However, the passage to which I refer does not address implications of the semiotic quality of concepts, but rather speaks of mistakes—ill-achievements—to be avoided.

7. J. Simon, *Philosophy of the Sign*.

8. See, e.g., Roland Barthes, *Elements of Semiology*, trans. Annette Lavers and Colin Smith (New York, 1973), p. 35 and passim.

9. A terminological explanation is perhaps in order here. What is meant here by "system reference" was formerly called "the one who signifies" (i.e., the sign-using subject) according to earlier linguistic usage. See, for instance, Novalis, "Philosophische Studien 1795/96 (Fichte-Studien)," in his *Werke, Tagebücher und Briefe Friedrich von Hardenbergs*, ed. Hans-Joachim Mähl and Richard Samuel, vol. 2 (Darmstadt, 1978), pp. 12ff. However, this left for the structural analysis of sign usage only the distinction of the sign and the signified. I prefer to follow the modern distinction between *signe* (sign), *signifiant* (signifier), and *signifié* (signified).

10. The passage in Heinz von Foerster's *Observing Systems* (Seaside, Calif., 1981) reads, "The environment contains no information; the environment is as it is" (p. 263).

11. See George Spencer-Brown, *Laws of Form* (reprint, New York, 1979). All subsequent quotations are taken from this reprint.

12. Here, I go beyond what can be found in Spencer-Brown. He simply states: "Call the space cloven by any distinction, together with the entire content of the space, the form of the distinction" (p. 4).

13. I will later deal with the endless discussions of the problem of reference triggered by such a suggestive simplification.

14. Spencer-Brown simply calls meaning, intending, signifying, and so forth "indication." He begins his investigation with the proposition: "We take as given the idea of distinction and the idea of indication, and that we cannot make an indication without drawing a distinction. We take, therefore, the form of distinction for the form" (p. 1). Note the proviso "therefore." Obviously, the distinction is precisely the form *par excellence* because the distinction as form implies the distinction between distinction and indication and thus contains itself as a necessary element. The processing of a form serves the purpose of unfolding the paradox.

15. Spencer-Brown accounts for this by using the injunction "draw a distinction" (p. 3) for choosing a distinction.

16. In the ninth Lowell lecture (1866), Peirce states, for example, "by an interpretant we mean a representation which represents that something is a representation of something else of which it is itself a representation." *Writings of Charles S. Peirce: A Chronological Edition*, ed. Max Frisch, Christian J. W. Kloesel, Edward C. Moore, et al., vol. 1 (Bloomington, 1982), p. 474.

17. See, e.g., Karl-Otto Apel's introduction to Charles S. Peirce, *Schriften II: Vom Pragmatismus zum Pragmatizismus* (Frankfurt a.M., 1970), p. 83.

18. For the way in which philosophers deal with antinomies in a corresponding and therefore "metaphysical" manner, see Nicholas Rescher, *The Strife of Systems: An Essay on the Grounds and Implications of Philosophical Diversity* (Pittsburgh, 1985).

19. Since Brunelleschi and Leonardo, "perspective" has been a standard notion for this state of affairs. It is a notion, however, that does not sufficiently clarify the structural distinction implicit in observing.

20. This corresponds to the recent terminology of systems theory, employed since von Foerster's *Observing Systems* and various investigations by Humberto Maturana. See, for example, Humberto R. Maturana and Francisco J. Varela, *Autopoiesis and Cognition: The Realization of the Living* (Dordrecht, 1980). See also Niklas Luhmann et al., eds., *Beobachter: Konvergenz der Erkenntnistheorie?* (Munich, 1990); and Niklas Luhmann, *Die Wissenschaft der Gesellschaft* (Frankfurt a.M., 1990), pp. 68–121.

21. Spencer-Brown, *Laws of Form*, p. 1.

22. To simplify the argument, I here omit from consideration the undoubtedly real possibility of there being a multiplicity of signifiers for the same signified, as is the case in the majority of languages with different words for the same object. In any event the argument does not change. If we were to include that possibility

in our argument, the use of signs would only be burdened with additional distinctions—for example, with the choice of language or an equivalent wording in which we want to say things.

23. Jakobson's critique of Saussure's principle of arbitrariness is thereby dissolved. See Roman Jakobson, "Zeichen und System der Sprache" (1962), in his *Semiotik: Ausgewählte Texte 1919–1982*, ed. Elmar Holenstein (Frankfurt a.M., 1988), pp. 427–39. Of course, it should not be assumed that Saussure himself developed the form analysis of the sign to the extent I have above. In that respect, and in view of the philological status of Saussure's text, Jakobson's critique is understandable. Yet Saussure after all provides the distinction between *signe* and *signifiant/signifié*, and this distinction can very well be read as the distinction between the unity (form) of the distinction and what is distinguished by the distinction.

24. In a different terminology one could also speak of *recursivity*, that is, of a process that constantly jumps ahead and backward to signs either not yet or no longer at the center. This can be observed in eye movements during reading. In the same theoretical context, Novalis employs the (Kantian) notion of the *schema*. A sign (used by the one who signifies, i.e., the subject) has to "stand in a *schematic* relationship to the signified" in order to be comprehensible (to another sign-user). A schema is the "unity of everything." It "stands in an interactive relationship with itself. Each thing in its given place is but what it is by virtue of the other things." This is found in Novalis's "Philosophische Studien 1795/96" (p. 14). There Novalis also affirms the principle of isolation within the self-referential circle of the sign and at the same time views it as the condition for comprehensibility (communicability, etc.).

25. When allegations such as that of arbitrariness, decisionism, and so forth arise, they are typically made in the context of a strategic defense of a theory, wherein the defender confronts the opponent with the presumed consequences of the latter's own theory. On this subject, see Stanley Fish's poignant *Doing What Comes Naturally: Change, Rhetoric, and the Practice of Theory in Literary and Legal Studies* (Oxford, 1989), pp. 7ff.

26. See John Lyons, *Semantics*, vol. 1 (Cambridge, Eng., 1977), pp. 305ff.

27. This development from an unreflected to a reflected understanding of signs, by the way, explains the close connection of linguistics and semiotics in earlier theoretical models that seemingly construe semiotics as a linguistic theory. Only more recent theories treat semiotics as a fundamental scientific discipline and linguistics as a case of applied semiotics. See Dean MacCannell and Juliet F. MacCannell, *The Time of the Sign: A Semiotic Interpretation of Modern Culture* (Bloomington, 1982); and Klaus Oehler, "Ist eine transzendentale Begründung der Semiotik möglich?" in K. Oehler, ed., *Zeichen und Realität: Akten des 3. semiotischen Kolloquiums Hamburg*, vol. 1 (Tübingen, 1984), pp. 45–59.

28. By making Peirce's insights resemble those of Karl-Otto Apel, Gerhard Schönrich reaches the opposite conclusion. See Gerhard Schönrich, *Zeichenhan-*

deln: Untersuchungen zum Begriff einer semiotischen Vernunft im Ausgang von Ch. S. Peirce (Frankfurt, 1990).

29. This has been especially true since Barthes's *L'empire des signes* (Paris, 1970); first American edition *Empire of Signs* (New York, 1982). The description of what is and remains a pure *signifiant* (signifier) then makes use of several metaphors such as those of writing and the body. However, the connections drawn in this world of mere *signifiants* remain associative. The effort was worthwhile, if only to demonstrate how far one might travel on this path.

30. This is particularly difficult in German, for purely stylistic reasons: time and again, the language tempts us to designate the signifier as the sign. See, for example, the usage in Schönrich's investigation. Not surprisingly, the simplification leaves behind a demand for a theory that could occasion metaphysical (Josef Simon) or transcendental-theoretical (Schönrich) solutions. The theory of form proposed here attempts to avoid those two stopgaps and, especially, the postmodern "signs are signs are signs" formula, even at the expense of stylistic concerns.

31. See Ranulph Glanville, "The Same Is Different," in Milan Zeleny, ed., *Autopoiesis: A Theory of Living Organization* (New York, 1981), pp. 252–62.

32. See MacCannell and MacCannell, *Time of the Sign*, especially "Second Semiotics: The Semiotic of Difference," pp. 152ff.

33. Ibid., p. 143, in reference to Peirce *and* Saussure.

34. Ibid., p. 149.

35. In reference to the specific case of art, see also Niklas Luhmann, "Weltkunst," in Niklas Luhmann, Frederick D. Bunsen, and Dirk Baecker, *Unbeobachtbare Welt: Über Kunst und Architektur* (Bielefeld, 1990), pp. 7–45.

36. This view is, however, held by MacCannell and MacCannell, *Time of the Sign*, pp. 146ff.

37. I present this suggestion without regard to a theology that, for its part, attempted to think of God as an entity without form (i.e., without any distinctions whatsoever) and to use the difference between God and the world in a variety of ways to give the world a form according to God's will. Despite such attempts, even the possibility of distinguishing between God and the world has been called into question by rigorously thinking theologians. As early a thinker as Nicholas of Cusa states in his *De Visione Dei* (book 9), "Extra te igitur, Domine, nihil est potest," with the result that everything in the world can be viewed as a "contracted" *visio Dei*. See Nicholas of Cusa, *Philosophisch-theologische Schriften*, ed. Leo Gabriel, vol. 3 (Vienna, 1967), p. 130. Consequently, the *contractio* pertains neither to God nor to the world but only to the self-observing observer.

38. A detailed discussion of Husserl and a critique of Derrida would be in order here. This would, however, lead us far afield. At any rate, Husserl's theory can be understood to center on the idea of the unity between consciousness and the phenomenon and thus of the unity between self-reference and external reference as constituent elements of any consciously aware activity. The form of this unity

would then be the intention, and the way in which meaning, as a phenomenon, is given can be gauged from intention. In that case, a theory of signs is not needed—a theory of signs with which Derrida quite effectively took issue. And likewise Husserl's transcendental-theoretical self-interpretation does not logically ensue from his central thesis. It is quite possible to imagine an empirical (autopoietic) consciousness operating intentionally.

39. For an interpretation of meaning in a sociologically oriented context, see Niklas Luhmann, "Sinn als Grundbegriff der Soziologie," in Jürgen Habermas and Niklas Luhmann, *Theorie der Gesellschaft oder Sozialtechnologie—Was leistet die Systemforschung?* (Frankfurt a.M., 1971), pp. 25–100; Niklas Luhmann, *Social Systems*, trans. John Bednarz, Jr., with Dirk Baecker (Stanford, Calif., 1995), pp. 59–102; and Niklas Luhmann, "Complexity and Meaning," in *Essays on Self-Reference* (New York, 1990), pp. 80–85.

40. This distinction can be traced back to Heider's psychological analyses of perception. See Fritz Heider, "Thing and Medium," *Psychological Issues* 1, no. 3 (1959): 1–34 [German 1926]. In the context of perception (as operation), Heider speaks not of "form" but of "thing." For the purpose of my proposed generalization of this distinction, I prefer the concept of form since it also expresses the paradoxical recurrence of forms within forms, that is, of distinctions within distinctions.

41. Spencer-Brown's calculus of form, to which we have referred several times, undertakes a reconstruction of Boolean arithmetic and algebra. Spencer-Brown explicitly rejects a characterization of his project as logic since it is about the processing of signs (in the sense of "tokens") and not about the processing of truth values.

42. This pertains in particular to the conception of self-referential systems, including such concepts as autopoiesis, self-organization, and operational closure.

43. MacCannell and MacCannell, *Time of the Sign*, pp. 1, 154ff.

44. Fragment no. 1954 according to the numbering in the Ewald Wasmuth's Novalis edition, *Fragmente*, vol. 2 (Heidelberg, 1957).

45. Here one cannot help but cite Ernst Cassirer, *The Philosophy of Symbolic Forms*, trans. Ralph Manheim, 3 vols. (New Haven, Conn., 1953–57 [German ed. 1923–29]).

46. See Walter Müri, *Symbolon: Wort- und sachgeschichtliche Studie* (Bern, 1931).

47. See Jan Assmann, *Ägypten: Theologie und Frömmigkeit einer frühen Hochkultur* (Stuttgart, 1984).

48. See also Goethe's rather playful wording in "Philostats Gemälde—Nachträgliches," in *Goethes Werke*, Weimar edition, reprint, sec. 1, vol. 49.1 (Munich, 1987), p. 142: A symbol "is the thing without being the thing and yet the thing; it is an image synthesized in the mirror of the mind and yet identical to the object."

Schorr, On the Analysis and Use of Form in Logic

1. See Kurt Gödel, *On Formally Undecidable Propositions of "Principia Mathematica" and Related Systems*, trans. B. Meltzer (New York, 1962).
2. For instance, George Spencer Brown, *Laws of Form*, 2d ed. (New York, 1972).
3. George Boole, *The Mathematical Analysis of Logic, Being an Essay Towards a Calculus of Deductive Reasoning* (Cambridge, Eng., 1847), p. 78; this, however, marks the beginning of a new kind of realism. See W. Baldamus, "Soziologie der formalen Logik," in N. Stehr and V. Meja, eds., *Wissenssoziologie* (Opladen, 1981), pp. 464–77.
4. Kant viewed and treated this discovery ("amphiboly") as a paradox of the ontological distinction between thought and Being, which presupposes as an option the distinction between transcendental and empirical. See Immanuel Kant, *Critique of Pure Reason*, trans. Norman Kemp Smith (New York, 1965), pp. 279–81 (B 322–24).
5. "Techniques" in a broad sense. See Niklas Luhmann, *Die Wissenschaft der Gesellschaft* (Frankfurt a.M., 1990), p. 197.
6. Boole, *Mathematical Analysis of Logic*, p. 3.
7. The axiom is based on the principles, that is, the laws of distributivity and commutativity, but also the laws of multiplication: "$x^* = x$," where $* = n$ (my insertion).
8. Boole, *Mathematical Analysis of Logic*, p. 18.
9. Gottlob Frege, *Begriffsschrift, eine dem Arithmetischen nachgebildete Formelsprache des reinen Denkens* (Halle, 1879); translated as *Conceptual Notation: A Formula Language of Pure Thought upon the Formula Language of Arithmetic*, in Gottlob Frege, *Conceptual Notation and Related Articles*, trans. and ed. Terrell W. Bynum (Oxford, 1972), pp. 101–203.
10. Gottlob Frege, *The Basic Laws of Arithmetic: Exposition of the System*, trans. and ed. Montgomery Furth (Berkeley, Calif., 1964), p. 23.
11. On writing as difference, see Jacques Derrida, *Of Grammatology*, trans. Gayatri Chakravorty Spivak (Baltimore, 1976). On writing as medium, see Luhmann, *Die Wissenschaft der Gesellschaft*.
12. In this respect, Frege's approach can be considered a first step toward the Turing machine.
13. On the use of this symbol, see Paul Lorenzen, *Formal Logic*, trans. Frederick J. Crosson (Dordrecht, 1965), pp. 30–31.
14. Frege, *Conceptual Notation*, p. 124.
15. In this respect, the analysis of form in Frege's *Begriffsschrift* (*Conceptual Notation*) of 1879 becomes dependent on decisions. But this does not imply that the openness of future states is recognized—an openness that would have to be "taken care of" by means of redundancy. When I discuss the de-ontologized functional orientation of logic I will return to this point.

16. See Frege, *Conceptual Notation*, p. 107.

17. Frege, *Basic Laws of Arithmetic*, p. 3. Here this kind of generality is understood as a shorthand for which Frege (in *Basic Laws of Arithmetic*) introduces the axiom of comprehension. Subsequent to Frege, mathematics' strategy for avoiding paradoxes would focus on this axiom.

18. Hence the "observation of observation," which is related to form production as an observer, still without being able to eliminate this distinction. See Frege's afterword in Gottlob Frege, *Grundgesetze der Arithmetik*, 2d ed. (Hildesheim, 1962), pp. 253ff.

19. Frege, *Basic Laws of Arithmetic*, p. 7.

20. See Niklas Luhmann, "The Paradox of Form," in this volume.

21. See the now-famous presentation given by David Hilbert in Paris in 1900, which, by enumerating 23 *singular* problems in mathematics, attests to the fact that there was already a different "mind set" at work, even though Hilbert and others still subscribed to the idea of deductive closedness. See Felix E. Browder, ed., *Mathematical Developments Arising from Hilbert Problems: Proceedings of the Symposium on Pure Mathematics*, 28th (Providence, R.I., 1976).

22. Russell's discovery had already prompted Frege to think about further specifying the "borderlines" between the conceptual level and the object level in order to rescue the idea of an "ontological" logic of either/or, but he could not bring himself to dissociate his analysis from traditional principles.

23. In the aftermath of the Gödel catastrophe, this has been further developed into a proof theory. See Kurt Schütte, *Proof Theory*, trans. J. N. Crossley (New York, 1977).

24. Spencer Brown, *Laws of Form*.

25. See, once again, Luhmann, "The Paradox of Form," in this volume.

26. That is, what constituted mathematics according to the "naïve" set-theoretical reformulations in the nineteenth century.

27. This then also applies indirectly to operative or other "effective" (including intuitionist) forms of logic when they focus, for instance, on "rescuing" the notion of a decision procedure for all kinds of formalism (calculi) in order to avoid the possible objection to always presupposing truth conditions (axioms). But for that, conditions (e.g., undeducible formulae or learnable semantics) must be provided first so one can operate "generally" with truth values. On the sophistication of this kind of logical investigation and related points, see Lorenzen, *Formal Logic*, pp. 61–72 (§9 Negation).

28. It must suffice to point to the central role of this concept in Luhmann's conception of science (see his *Die Wissenschaft der Gesellschaft*), which I follow here.

29. Luhmann distinguishes between redundancy and variety: variety, in contrast to the redundancy, denotes the number and diversity of events, a definition that reflects the function-specific view on selectivity. I do not want to delve further into this distinction here, although it would allow us, at least from a socio-

logical perspective, to shed light on the abbreviation technique of operations that plays a role in Frege's abstractions. For more details, see Luhmann, *Die Wissenschaft der Gesellschaft*, pp. 436ff.

30. This question is underscored by the difficulties raised when introducing negation into a logically interpretable semiotic context that is no longer bound to an ontological understanding of the *tertium non datur*. I will briefly return to this point later.

31. On the concept of conditioning, see Niklas Luhmann, *Social Systems*, trans. John Bednarz, Jr., with Dirk Baecker (Stanford, Calif., 1995), p. 23.

32. According to Luhmann, *Social Systems*, p. 363.

33. One might consider, for example, the precarious relationship between pedagogy and the science of education.

34. See Lorenzen, *Formal Logic*, p. 71. He makes the extraordinary remark that "classical logic" itself has never led to contradictions.

Esposito, Two-Sided Forms in Language

1. See Niklas Luhmann, "Sign as Form," in this volume.
2. See George Spencer Brown, *Laws of Form* (London, 1969).
3. Spencer-Brown avoids the idea of a starting point free of presuppositions (instead he introduces the idea of "re-entry") and thereby becomes free to construct a calculus without initial postulates. The most important point is operationality. For the logical intricacies of the calculus of indications, see Elena Esposito, *L'operazione di osservazione: Construttivismo e teoria dei sistemi sociali* (Milan, 1992).
4. See Ranulph Glanville, "Beyond the Boundaries," in R. F. Ericson, ed., *Improving the Human Condition: Quality and Stability in Social Systems* (Louisville, Ky., 1979), pp. 70–74; Ranulph Glanville and Francisco J. Varela, "Your Inside Is Out and Your Outside Is In," in G. E. Lasker, ed., *International Congress on Applied Systems Research and Cybernetics*, vol. 6 (New York, 1981), pp. 638–41; and Paul Cull and William Frank, "Flaws of Form," *International Journal of General Systems* 5 (1979): 201–11.
5. See Francisco J. Varela, "A Calculus for Self-Reference," *International Journal of General Systems* 2 (1975): 5–24.
6. Spencer Brown, *Laws of Form*, pp. 7 and 69.
7. This applies also to the case of self-observation in which the system, although observing itself, does not observe the ongoing operation.
8. Humberto R. Maturana employs this expression to signify those relations independent of the systemic autopoiesis in which elements of a system participate. See Maturana, "The Biological Foundations of Self-Consciousness and the Physical Domain of Existence," in Niklas Luhmann et al., eds., *Beobachter: Konvergenz der Erkenntnistheorien?* (Munich, 1990), pp. 47–117, quotation from p. 78. This expression had previously been used by G. Sommerhoff: "Two variables are

orthogonal, if the value of the one at any instant of time does not determine the value of the other *for the same instant*." G. Sommerhoff, "The Abstract Characteristics of Living Systems," in F. E. Emery, ed., *Systems Thinking: Selected Readings* (London, 1969), pp. 147–202, quotation from p. 155.

9. In this essay, I will treat the issue of form entirely on the level of observation. In other words, I will assume that form, as a two-sided form, only makes sense for an observing system. The premises of this assertion cannot be detailed here. For a more elaborated treatment, see Esposito, *L'operazione di osservazione*.

10. This distinction in turn differs from the one between the "form" and "substance" of an expression, according to which apple the object does not consist of five letters and the word "apple" is not sweet. More on this later. On the form/substance distinction, see Louis Hjelmslev, *Prolegomena to a Theory of Language*, trans. Francis J. Whitfield (Madison, Wis., 1961).

11. See Gotthard Günther, "Cognition and Volition: A Contribution to a Cybernetic Theory of Subjectivity," in his *Beiträge zur Grundlegung einer operationsfähigen Dialektik*, vol. 2 (Hamburg, 1979), pp. 203–40.

12. Our investigation into the form of language produces a conception that is essentially similar to Wilhelm von Humboldt's classic definition. Humboldt related the notion of linguistic form to the distinction between grammar and lexicon, and primarily conceived of it such that language gives form to matter (and matter as matter only exists in regard to language). See Wilhelm von Humboldt, *On Language: The Diversity of Human Language-Structure and its Influence on the Mental Development of Mankind*, trans. Peter Heath (Cambridge, Eng., 1988).

13. See Luhmann, "Sign as Form," in this volume. For the moment, I will postpone addressing the much-contested issue of the relationship between linguistic and nonlinguistic signs. Here I confine myself to the assumption that propositions about signs in general pertain to linguistic signs as well.

14. The notation s/S that I will employ hereinafter to represent the signifier/signified distinction stems from the Lacanian school. See Jacques Lacan, *Écrits* (Paris, 1966); Anika Lemaire, *Jacques Lacan* (Brussels, 1970).

15. Ferdinand de Saussure, *Course in General Linguistics*, ed. Charles Bally and Albert Sechehaye, with the collaboration of Albert Riedlinger, trans. Wade Baskin (London, 1974), p. 103 (pt. 2, chap. 2).

16. "If we then say that a message has 'meaning' or is 'about' some referent, what we mean is that there is a larger universe of relevance consisting of message-plus-referent, and that redundancy or pattern or predictability is introduced into this universe by the message." So says Gregory Bateson, *Steps to an Ecology of Mind: Collected Essays in Anthropology, Psychiatry, Evolution and Epistemology* (New York, 1972), p. 413.

17. "Identity" as defined by Niklas Luhmann, "Identität—Was oder wie?" in his *Soziologische Aufklärung 5: Konstruktivistische Perspektiven* (Opladen, 1990), pp. 14–30.

18. The play with the magic of words in myths (see, for example, Walter J. Ong, *Orality and Literacy: The Technologizing of the Word* [New York, 1982]) or the inclination of children to equate the characteristics of names with those of the signified objects confirms the natural tendency to establish "motivated" relations between signifier and signified. Even Gaston Bachelard speaks of words' "internal oneirism," owing to which, for instance, the gender of a word is linked to its meaning: Depending on the linguistic gender, masculine or feminine connotations are even attributed to genderless objects. See Gaston Bachelard, *The Poetics of Reverie*, trans. Daniel Russell (New York, 1969), p. 35.

19. See André Martinet, *La linguistique synchronique*, 4th ed. (Paris, 1974), chap. 1.

20. It is a glossematic isomorphism, that is, a parallelism between the level of expression and the content level. See Hjelmslev, *Prolegomena to a Theory of Language*.

21. The position I take here thus contradicts the opinion that the particular focus on language in semiotic research is a consequence of what might be called the "verbocentric fallacy." Roland Barthes, for example, suggests that one ought to adopt a more flexible stance that grants "sign systems whose substance is not verbal" the same rank as language. See Roland Barthes, *Elements of Semiology*, trans. Annette Lavers and Colin Smith (London, 1967). I maintain, by contrast, that even if the concept of a sign is more comprehensive than that of a linguistic sign, there would be no autonomy of nonlinguistic signs if linguistic means were not available. Consequently, neither would there be other "sign systems"—the reason being that two-sided forms are necessary.

22. Saussure, *Course in General Linguistics*, p. 69.

23. Ibid., p. 67.

24. Emile Benveniste, "The Nature of the Linguistic Sign," in his *Problems in General Linguistics*, trans. Mary Elizabeth Meek (Coral Gables, Fla., 1971), pp. 43-48, quotation from p. 45.

25. Fashion therefore would be an example of an arbitrary sign system; language, on the other hand, would not. See Barthes, *Elements of Semiology*, p. 51.

26. Saussure, *Course in General Linguistics*, p. 69.

27. Ibid. Also see Barthes, *Elements of Semiology*, p. 50. Regarding this, Umberto Eco has distinguished between a *ratio facilis* and a *ratio difficilis*. See Umberto Eco, *A Theory of Semiotics* (Bloomington, 1979), pp. 183-84. In the case of the *ratio difficilis* the difficulty of signification not surprisingly lies mainly in the problem of repeating the expression.

28. Saussure, *Course in General Linguistics*, p. 74.

29. Ibid., p. 73.

30. Ibid., p. 75.

31. For example, see Ranulph Glanville, "Distinguished and Exact Lies," in R. Trappl, ed., *Cybernetics and Systems Research*, vol. 2 (Amsterdam, 1984), pp. 655-62; or Humberto R. Maturana and Francisco J. Varela, *The Tree of Knowl-*

edge: The Biological Roots of Human Understanding, trans. Robert Paolucci, rev. ed. (Boston, 1992).

32. John Lyons, *Language and Linguistics: An Introduction* (Cambridge, Eng., 1981), p. 19.

33. See Niklas Luhmann, *Die Wissenschaft der Gesellschaft* (Frankfurt a.M., 1990), pp. 100 and 175.

34. Cited in Barthes, *Elements of Semiology*, p. 51.

35. A closed system, says Bateson, "will generate a non-random response to a random event at that position in the circuit at which the random event occurred." Bateson, *Steps to an Ecology of Mind*, p. 410.

36. See Saussure, *Course in General Linguistics*, pp. 161–62.

37. An analog computer, for instance, establishes that "a certain intensity of current x denotes a physical size y, and that the denotative relation is based on a proportional one.... In other words the proportion depends upon the fact that if size 10 corresponds to intensity 1, size 20 will correspond to intensity 2, and so on." Eco, *A Theory of Semiotics*, pp. 200–201.

38. The use of binary digits in the realm of computers is based on defining the information of a message as the smallest number of binary decisions that enables the receiver to reconstruct the content of a message. See Roman Jakobson, "Linguistics and Communication Theory," in *Proceedings of Symposia in Applied Mathematics, Vol. XII: Structure of Language and Its Mathematical Aspects* (Providence, R.I., 1961), pp. 245–52.

39. Barthes, *Elements of Semiology*, p. 54.

40. See Nikolai Sergeevich Trubetskoi, *Principles of Phonology*, trans. Christiane A. M. Baltaxe (Berkeley, Calif., 1969); Roman Jakobson and Morris Halle, "Phonology and Phonetics," in Roman Jakobson, *Selected Writings*, vol. 1 (The Hague, 1962), pp. 464–504.

41. Roman Jakobson, "The Zero Sign" (1939), in his *Russian and Slavic Grammar: Studies 1931–1981*, ed. Linda R. Waugh and Morris Halle (Berlin, 1984), pp. 151–60.

42. Barthes, *Elements of Semiology*, p. 73.

43. "Language is satisfied with the opposition between something and nothing." Saussure, *Course in General Linguistics*, p. 86.

44. Ibid., pp. 110–22.

45. Ibid., p. 117.

46. Ibid., p. 120.

47. Ibid., p. 122. The property of *discreteness*, often considered as probably one of the distinguishing features of language, is also based on this distinction. See Giulio C. Lepschy, *A Survey of Structural Linguistics* (London, 1970), pp. 32–35; Charles F. Hockett and Stuart A. Altman, "A Note on Design Features," in T. A. Sebeok, ed., *Animal Communication: Techniques of Study and Results of Research* (Bloomington, 1968), pp. 61–72, especially pp. 63ff. In languages, identity of form is an all-or-nothing issue and not one of more-or-less. If the vowels in

"palo" and "polo" are replaced with any sound between "a" and "o," no third vowel or no new word is attained but instead something unrecognizable (or not recognizable as either of the two). See Lyons, *Language and Linguistics*, pp. 21ff. What counts is therefore the articulation of the distinctions and not the "phonetic substance."

48. Saussure, *Course in General Linguistics*, p. 118.

49. Ibid.

50. Barthes, *Elements of Semiology*, p. 57.

51. This is essentially also the question behind the issue of *pertinence*. See Luis J. Prieto, "La découverte de phonème: Interprétation épistémologique," *La Pensée* 148 (1969): 35–53. With the "discovery of the phoneme" leading to the uncovering of the "empirical fallacy" (p. 51), the problem of linguistics no longer consists in explaining why the speaking subject disregards some sound properties that it could realize physically, but in explaining the fact that it does recognize specific properties in a sound at all, "since obviously the fact that the sound exhibits those properties could not suffice as an explanation any more" (p. 52). Thus the problem lies in "the distinctions the subject *makes*," and not in those it "does not make": The issue then becomes investigating the "pertinence" of particular distinctions "since an object can be the reference point for an infinity of distinctions" (p. 52).

52. See Martinet, *La linguistique synchronique*, chap. 1.

53. See ibid., p. 32.

54. Ibid.

55. See John Lyons, *Semantics*, vol. 1 (Cambridge, Eng., 1977), p. 76.

56. See Niklas Luhmann, *Social Systems*, trans. John Bednarz, Jr., with Dirk Baecker (Stanford, Calif., 1995), p. 443; and Claudio Baraldi, Giancarlo Corsi, and Elena Esposito, *Luhmann in Glossario: I concetti fondamentali della teoria dei sistemi sociali* (Milan, 1996), pp. 58–59.

57. Martinet, *La linguistique synchronique*, p. 13.

58. When the segmentation (articulation) into phonemes is transferred into another medium—as in the case of the phonetic writing system—its distinguishing function is indeed not affected. The issue of the plurality of media will be addressed later.

59. Martinet, *La linguistique synchronique*, p. 34.

60. Ibid., p. 14.

61. Ibid., p. 34.

62. This is connected to the "law" of absolute regularity of phonetic change, which follows "sound laws without exceptions" characteristic of the "neogrammarian" approach. See Hermann Osthoff and Karl Brugmann, *Morphologische Untersuchungen auf dem Gebiet der indogermanischen Sprachen*, vol. 1 (Leipzig, 1878); Lyons, *Language and Linguistics*, pp. 192ff.; and Robert Henry Robins, *A Short History of Linguistics*, 3d ed. (London, 1990), pp. 201–11.

63. See Barthes, *Elements of Semiology*, pp. 72–73.

64. See Luhmann, *Social Systems*, p. 285; and Baraldi, Corsi, and Esposito, *Luhmann in Glossario*, p. 194.

65. In Saussurian terminology, the plane of *langue*.

66. Saussure, *Course in General Linguistics*, pp. 122–27. For each relationship, the following terms are equivalent: (1) "relations" (Hjelmslev), "contiguities" (Jakobson), "contrasts" (Martinet), and "syntagmatic connections" (Barthes); (2) "correlations" (Hjelmslev), "similarities" (Jakobson), "oppositions" (Martinet), and "systematic connections" (Barthes).

67. Paradigmatic relations were originally called "associative relations."

68. Hence, Peirce's "infinite semeiosis"; Charles Sanders Peirce, *Collected Papers*, ed. C. Hartshorne, P. Weiss, and A. W. Burks, vol. 1 (Cambridge, Mass., 1965), p. 339. See also Eco's "model Q" (*A Theory of Semiotics*, pp. 121–25), in which every sign is defined in conjunction with the entirety of other signs.

69. What makes the organization of language on two axes concretely significant for the functioning of language and the production of forms in general has been demonstrated, for example, in Jakobson's investigations on two different forms of aphasia. See Roman Jakobson and Morris Halle, *Fundamentals of Language* (The Hague, 1956). The first form of aphasia is marked by a "disorder of similarity" that prevents the attribution of meaning to words outside a given context. Distinctions on the paradigmatic axis cannot be appropriately performed, so the aphasiac operates mainly metonymically. The second form of aphasia, by contrast, is a "disorder of contiguity" that affects the ability to combine linguistic units into sentences; it leads to agrammatism. In this case, distinctions on the syntagmatic axis cannot be carried out, so the aphasiac operates in "quasi-metaphorical expressions." This finding is relevant for several reasons, among them that it empirically validates the conclusion that the inability to handle i/d distinctions leads to the inability to signify at all (i.e., to carry out s/e distinctions). Those suffering from aphasia are often incapable of giving an object its proper name.

70. Barthes, *Elements of Semiology*, p. 65.

71. On purely acoustic grounds these contexts are very different; one only has to think of the differences in sound timbre, pronunciation, intensity, and so forth.

72. Barthes, *Elements of Semiology*, p. 66.

73. Hjelmslev, *Prolegomena to a Theory of Language*, pp. 68ff.; Lepschy, *A Survey of Structural Linguistics*, pp. 117–18.

74. See Fritz Heider, "Thing and Medium," *Psychological Issues* 1, no. 3 (1959): 1–34 [German 1926]; Niklas Luhmann, "Das Medium der Kunst," *Delfin* 4 (1986): 6–15; Niklas Luhmann, "Gleichzeitigkeit und Synchronization," in his *Soziologische Aufklärung 5: Konstruktivistische Perspektiven* (Opladen, 1990), pp. 95–130; Luhmann, *Die Wissenschaft der Gesellschaft*; and Baraldi, Corsi, and Esposito, *Luhmann in Glossario*, pp. 118–19.

75. Referring to cases of sense perception (e.g., hearing the ticking of a watch), Fritz Heider (in "Thing and Medium") makes a similar point: "The vibrations of the air ... are not represented phenomenally.... They spend them-

selves, so to speak, in the process of mediation so that we believe that we hear the ticking directly. In this case we are ordinarily not aware that mediation exists. The mediation of light waves is of the same nature. We do not perceive light waves as things that touch our eyes and refer to something else. We seem to see the mediated object directly" (p. 2).

76. "Yet at whatever time and in whatever way we speak a language, language itself never has the floor." Martin Heidegger, *On the Way to Language*, trans. Peter D. Hertz (New York, 1971), p. 59.

77. See, for example, Niklas Luhmann, "The Form of Writing," *Stanford Literature Review* 9 (1992): 2-42.

78. Luhmann, "Das Medium der Kunst," p. 6.

79. Ibid.

80. This does not change the fact that every form performs a selection: There are fewer phonemes than there are words, yet the number of possible combinations of phonemes is higher than the number of words in a particular language.

81. See Niklas Luhmann, "Sozialisation und Erziehung," in W. Rotthaus, ed., *Erziehung und Therapie in systemischer Sicht* (Dortmund, 1987), pp. 77-86; and Niklas Luhmann, "How Can the Mind Participate in Communication," in Hans Ulrich Gumbrecht and K. Ludwig Pfeiffer, eds., *Materialities of Communication*, trans. William Whobrey (Stanford, Calif., 1994), pp. 371-87.

82. Luis J. Prieto, *Messages et signaux* (Paris, 1966), p. 117.

Baecker, The Form Game

1. Italo Calvino, *Six Memos for the Next Millennium: The Charles Eliot Norton Lectures 1985-86*, trans. Patrick Creagh (New York, 1993), p. 57.

2. George Spencer Brown, *Laws of Form* (London, 1969). Page references to this edition will be included in my text.

3. Heinz von Foerster, "Objects: Tokens for (Eigen-)Behaviors," in his *Observing Systems* (Seaside, Calif., 1981), p. 273; Niklas Luhmann, "Identität—Was oder wie?" in his *Soziologische Aufklärung 5: Konstruktivistische Perspektiven* (Opladen, 1990), pp. 14-30.

4. Matthias Varga von Kibéd, "Wittgenstein und Spencer Brown," in Paul Weingartner and Gerhard Schurz, eds., *Philosophie der Naturwissenschaften: Akten des 13. Internationalen Wittgenstein Symposiums* (Vienna, 1989), pp. 402-6, quotation from p. 406.

5. Francisco J. Varela, Evan Thompson, and Eleanor Rosch, *The Embodied Mind: Cognitive Science and Human Experience* (Cambridge, Mass., 1991), pp. 59-63 and 219-26; Carlos Castaneda, *Journey to Ixtlan: The Lessons of Don Juan* (Harmondsworth, Eng., 1973).

6. Jacques Lacan, "Subversion du sujet et dialectique du désir dans l'inconscient freudien," in his *Ecrits II* (Paris, 1971), pp. 151-91, quotation from p. 161.

7. Talcott Parsons et al., "Some Fundamental Categories of the Theory of Ac-

tion: A General Statement," in Talcott Parsons and Edward A. Shils, eds., *Toward a General Theory of Action* (Cambridge, Mass., 1951), pp. 3–29, quotation from pp. 15–16.

8. Niklas Luhmann, *Social Systems*, trans. John Bednarz, Jr., with Dirk Baecker (Stanford, Calif., 1995), pp. 104–5.

9. Niklas Luhmann, "Operational Closure and Structural Coupling: The Differentiation of the Legal System," *Cardozo Law Review* 13 (1992): 1419–41, quotation from p. 1423.

10. Luhmann, *Social Systems*, p. 106.

11. Johan Huizinga, *Homo Ludens: A Study of the Play-Element in Culture*, trans. R. F. C. Hull (Boston, 1955); Roger Caillois, *Man, Play, and Games*, trans. Meyer Barash (New York, 1961); Eugen Fink, *Spiel als Weltsymbol* (Stuttgart, 1960); Hans-Georg Gadamer, *Truth and Method*, 2d rev. ed., trans. Joel Weinsheimer and Donald G. Marshall (New York, 1989), pp. 101–34; Jacques Derrida, "Structure, Sign, and Play in the Discourse of the Human Sciences," in his *Writing and Difference*, trans. Allan Bass (Chicago, 1978), pp. 278–93; Keiji Nishitani, *Religion and Nothingness*, trans. Jan Van Bragt (Berkeley, Calif., 1982), pp. 218–85.

12. Ludwig Wittgenstein, *Philosophical Investigations*, trans. G. E. M. Anscombe, 3d ed. (Englewood Cliffs, N.J., 1958), §7 (p. 5e).

13. Ibid., §31 (p. 15e).

14. Oskar Morgenstern, "Die Theorie der Spiele und des wirtschaftlichen Verhaltens," *Jahrbuch für Sozialwissenschaft* 1 (1950): 113–39.

15. John von Neumann and Oskar Morgenstern, *Theory of Games and Economic Behavior* (Princeton, N.J., 1953), p. 49.

16. Stephen Miller, "Ends, Means, and Galumphing: Some Leitmotifs of Play," *American Anthropologist* 75 (1973): 87–98.

17. Gregory Bateson, *Mind and Nature: A Necessary Unity* (New York, 1979), p. 139.

18. Gregory Bateson, "The Message 'This Is a Play,'" in Bertram Schaffner, ed., *Group Processes: Transactions of the Second Conference* (New York, 1956), pp. 145–242, quotations from pp. 148 and 150.

19. Gregory Bateson, "A Theory of Play and Fantasy," in his *Steps to an Ecology of Mind: Collected Essays on Anthropology, Psychiatry, Evolution, and Epistemology* (New York, 1972), pp. 177–200.

20. Bateson, *Steps to an Ecology of the Mind*, p. 185.

21. Fink, *Spiel als Weltsymbol*, p. 78.

22. Bateson, *Steps to an Ecology of the Mind*, pp. 179–81.

23. Huizinga, *Homo Ludens*, pp. 7–18, especially p. 13.

Hutter, The Early Form of Money

1. Note that I am here conceiving of payments as forms integrated into the

monetary medium. This diverges from the opinion that takes prices to be such forms within the monetary system. See Niklas Luhmann, "Das Moderne der modernen Gesellschaft," in Wolfgang Zapf, ed., *Die Modernisierung moderner Gesellschaften: Verhandlungen des 25. Deutschen Soziologentages in Frankfurt am Main 1990* (Frankfurt a.M., 1991), pp. 87–108, quotation from p. 99. However, prices are only signals of expectations by participants in the economy. Only in payments are prices actualized and thereby reproduced in the system.

2. See M. Ernest Babelon, *Les origines de la monnaie* (Paris, 1897), p. 106; and Miriam S. Balmuth, "Origins of Coinage," in *A Survey of Numismatic Research 1966–1971* (New York, 1973), pp. 27–34, especially pp. 30–31. Electrum, also known as "white gold," is a naturally occurring alloy of gold and silver. From a semiotic point of view, the use of metals is in itself a remarkable phenomenon. Signs are not created arbitrarily; they are not introduced into a free space. They are distinguished within an already existing environment that seems to be "close-packed," thus completely occupied by other signs. This means in the case of metals that the semiotic quality of specific metals is only established in a slow process alongside their material property. And the issue is always meanings addressed to an "other," be it a deity to whom one brings a sacrifice or a person for whom one adorns oneself.

3. See Colin M. Kraay, *Archaic and Classical Greek Coins* (London, 1976), p. 28.

4. See Fritz Heichelheim, *Wirtschaftsgeschichte des Altertums* (Leiden, 1938), and Fritz Heichelheim, "Geld- und Münzgeschichte I: Anfänge und Antike," in *Handwörterbuch der Sozialwissenschaften* (Tübingen, 1965).

5. Norman Davis, *Greek Coins and Cities* (London, 1961), p. 21.

6. The ambiguity of the German word *Versprechen* encapsulates the ambiguity of the observation. In one sense of *Versprechen*, someone promises an event that lies in the future. Current events are thus connected to future events, which have not yet happened. Though they are invented, they are possible according to the logic of the communication of payment. In the other sense of *Versprechen*, someone mistakenly attributes an observation to an earlier event that never happened. Hence he invents past events, which can then lead to a future consistent with those events. The triggering moment, however, can also lie with the one who offers the sign: he makes a mistake, he makes a slip of tongue, and yet the sign is accepted because the receiver does not perceive it as a mistake but contrives a suitable meaning.

7. It is interesting to note that the varying metallic content of electrum was of great importance even as early as the punchmarked coins. Accordingly, the punchmarks occurred predominantly on electrum pieces.

8. On dating issues, see below, note 24.

9. This is evidenced by the lack of traces of two hammering operations, which would be visible on the other side of the coin. See Lieselotte Weidauer, *Probleme der frühen Elektronprägung* (Fribourg, 1975).

10. The numismatic discourse usually refers to the side that rests on top of the anvil as the coin's obverse. This is certainly justified once this side is imprinted with images and hence becomes the more significant side of the coin. Regarding the pieces considered here, however, the reverse was probably still the more important side and thus functioned like the image-bearing obverse of later coins.

11. See Heichelheim, *Wirtschaftsgeschichte des Altertums*.

12. See Bernhard Laum, *Heiliges Geld* (Tübingen, 1924). From a systems-theoretical viewpoint it is worth noting that the distinction between gold and silver can provide a material analogy for the distinction between day and night. What is decisive here is the excluded third, that is, the world beyond the visible world of day and night. This fundamental closure can be represented by gold and silver. In my view, another property of metals carries communicative meaning as well: their reflectiveness. Metals throw back the images of the external world and thereby point to the fact that behind their surface another world begins.

13. See especially the account in Laum, *Heiliges Geld*.

14. It suffices to point out that, apart from the Pactolus River near Sardis, electrum from fluvial sand has only been discovered thus far at three other sites on the globe. See Balmuth, "Origins of Coinage," p. 31.

15. Sources on the non-European history of money are scarce. However, it appears that India's autochthonous monetary evolution never went beyond the form of marked metal pieces (i.e., form 1). See Bernhard Laum, "Münzwesen," in *Handwörterbuch der Sozialwissenschaften* (Jena, 1925); Pran Nath, *Tausch und Geld in Altindien* (Leipzig, 1924); and Babelon, *Origines de la monnaie*. The Chinese communication of payment up to the modern era also used precious-metal coins only occasionally. See Lien-Sheng Yang, *Money and Credit in China: A Short History* (Cambridge, Mass., 1952).

16. "It was the monopoly in stamped pieces of electrum that brought the first tyrant to the king's palace and placed him on the throne." Peter N. Ure, *The Origin of Tyranny* (Cambridge, Eng., 1922), p. 152. Ure's study gives the most detailed account of the historical episode we are presenting here. Although not all of his dates are still tenable today, the order of key events that he established does concur with our chronology of coin forms.

17. The notion of "tyranny" originally meant nothing more than the exercise of power by a ruler who is not legitimized transcendentally but in effect biologically. Only much later does this term take on the connotation of cruel and unjust rule.

18. See Ure, *Origin of Tyranny*, p. 143. Kraay, in *Archaic and Classical Greek Coins*, suggests that pictorial signs were initially related to the person in power, which meant that they had to be restruck when power changed hands.

19. The temporal priority of the lion's head over all other signs, however, is not completely evidenced by archeological findings. It is also conceivable that private houses experimented with pictorial signs. The religious connotation of the

emblem, however, constitutes the evolutionary difference. See Weidauer, *Probleme der frühen Elektronprägung*.

20. "The totem (consisting of animals, plants, implements, weapons) is the outward expression of a 'mystic participation' . . . ; he who carries the totemic symbol stands in magic communion with the totemic community; the symbol is a sanctioning of tribal membership." Laum, *Heiliges Geld*, p. 140.

21. It is remarkable that in this early phase written signs did not play any role. Though their use in marking goods had been established for a long time, their readability was far lower than that of pictorial signs. See Denise Schmandt-Besserath, "An Ancient Token System: The Precursor to Numerals and Writing," *Archaeology* 39 (1986): 32–39.

22. See Fritz Heichelheim, "Die Ausbreitung der Münzgeldwirtschaft und der Wirtschaftsstil im archaischen Griechenland," *Schmollers Jahrbuch* (1931): 37–62, quotation from pp. 42–43.

23. The effect of the fluctuating metallic content of the electrum coins is worth noting. While the weight of the discovered pieces remains constant within a 3 percent range, the gold content varies between 27 percent and 52 percent. See Kraay, *Archaic and Classical Greek Coins*, p. 28; and Josef Dobretsberger, "Vom Ursprung des Münzgeldes," *Finanzarchiv* (1961): 60–70, quotation from pp. 68–69.

Thus the stamping of the coin only guaranteed its weight. Since it became readily observable that the metallic content of the electrum pieces could change, the difference between signified value and signifying metal became apparent as well.

24. Archeological evidence figures, of course, as the prime source for dating coins. Interpretations are controversial, however. The key evidence is a single hoard left as a building sacrifice in the foundation of the Artemis temple at Ephesus and excavated in 1908. It contains 81 electrum coins, 9 "pre-coins," and several precious-metal objects. See Weidauer, *Probleme der frühen Elektronprägung*, and Kraay, *Archaic and Classical Greek Coins*. One of the coins carries an inscription that for a long time was attributed to the Lydian king Alyattes. Alyattes' rule began in 610 B.C. From the assumption that coins were used no longer than one generation, it was concluded that the earliest coins date to roughly 630 B.C. This line of reasoning goes back to E. S. G. Robinson, "The Date of the Earliest Coins," *Numismatic Chronicle and Journal of the Royal Numismatic Society* (1956): 1–8. By contrast, a more recent study—Weidauer, *Probleme der frühen Elektronprägung*—develops a different chronology, which we use in this text. The major arguments are the following: (1) It is improbable that the inscription can be positively linked to Alyattes; (2) there are Assyrian sources, composed no later than 626 B.C. given Ashurbanipal's death in that year, which report an attack on Ephesus by the Cimmerii in the course of which the Artemis temple was destroyed; (3) coins, and likewise the pre-coin objects, were in use much longer than one generation; (4) strong stylistic similarities can be demonstrated between art

objects from the first half of the seventh century and coin types, especially the lion heads.

25. This finding contradicts the interpretation of the emergence of coins common to the traditional theories of money. According to this understanding, coins once "invented" by a ruler's decree or by chance disseminate quickly in the economic sector since the benefits of this medium are supposed to be easily recognizable to the trading individuals.

26. This shows again the effect of the dissemination of money by means of signs: the coin value diverges from the material value, be it through counterfeit or certain reminting procedures. It is precisely "the false," that is to say, the deviation between the credibility of the material value and the face value of the coin sign, that makes it possible to trust in signs and thus to increase the amount of mintable coins.

27. According to a tradition that goes back to Herodotus, the introduction of silver currency is ascribed to King Pheidon of Argos. If this were true, the coinage would date to approximately 650 B.C. This, however, is not supported by existing coin findings. See Kraay, *Archaic and Classical Greek Coins*, pp. 41–42. Herodotus also reports that Pheidon confiscated the archaic spits that had been used as means of payment. Again, the deliberate act is probably a literary fiction. Yet it is nonetheless striking that money in the form of implements still was actively used in that era.

28. It is noteworthy that the Aeginetan rather than the Phoenician merchant clan succeeded in using coins in their trade with different societies. Even in this case, however, the coins were most likely employed for internal commerce among merchants. The initially conservative situation of the various people with whom the merchants traded, one can suspect, only allowed a slow penetration of the monetary code.

29. Given the value ratio 1 to 15, which remained stable over a long period of time, the Aeginetan drachma was calibrated in such a way that ten silver pieces equaled the value of one gold piece. See Ure, *Origin of Tyranny*, p. 172.

30. There was one transitional step in which both sides were stamped with animal images. Such coins appeared first in Corinth and subsequently in other cities as well. See Laum, "Münzwesen," p. 678.

31. The following account is primarily based on Kraay, *Archaic and Classical Greek Coins*.

32. Though the attribution to Hippias is controversial, Kraay puts forth convincing arguments.

33. Archaic owl coins were struck almost exclusively in tetradrachms, thus in large units of weight. They were discovered in coin finds between Sicily and Egypt.

34. Kraay, *Archaic and Classical Greek Coins*, p. 61.

35. See Heichelheim, "Die Ausbreitung der Münzgeldwirtschaft," p. 46. He speaks of a "sudden" expansion.

36. For the empire of Alexander the Great (i.e., around 330 B.C.), 162 minting sites and the production of 1.4 billion drachmae are reported. See Peter R. Franke and Max Hirmer, *Die griechische Münze* (Munich, 1972), p. 125.

37. Dirk Baecker especially experiments with this oscillation between private and public markings. See his works *Womit handeln Banken?* (Frankfurt, 1991) and "Die Metamorphosen des Geldes," *Delfin* 14 (1990): 17–26.

38. See C. H. V. Sutherland, "Corn and Coin: A Note on Greek Commercial Monopolies," *American Journal of Philology* 64 (1942): 129–47.

39. On the gradual merging of the concepts of interest on borrowed cattle, seeds, and metal see Heichelheim, *Wirtschaftsgeschichte des Altertums*, p. 224.

40. See Weidauer, *Probleme der frühen Elektronprägung*.

41. "On Alexander the Great's silver tetradrachm, the individual features of the king are inscribed into the head of the youthful Heracles adorned with the skin of the Nemean Lion. It cannot be ignored that the ambiguity of the images was indeed intended." Karl Christ, "Die Griechen und ihr Geld," *Saeculum* (1964): 214–29, quotation from pp. 221–22.

The chronology, however, is incomplete. Significantly earlier, heads of Persian satraps had already appeared on coins. They were modeled after Athena's head on the owl coin. See Kraay, *Probleme der frühen Elektronprägung*, and Franke and Hirmer, *Die griechische Münze*, p. 123. Yet it is true that those coin types were not part of a continuing tradition. From an evolutionary viewpoint, they remained irrelevant for the developments to follow.

42. "Back then, Alexander's silver currency ruled everywhere coins were minted under Greek influence, that is, from the Pillars of Hercules in the West to the Indus in the East. The Roman dinar seems to be the direct continuation of the Alexandrian drachma." Laum, "Münzwesen," p. 679.

43. It is, however, noteworthy that the change in the design of bank notes is frequently perceived as a transition to play money. In a new form the artificial nature of the value form is noticeable until habit blinds our view.

44. See Baecker, *Womit handeln Banken?* Crump's theory of money, though not based on a systems-theoretical approach, postulates in a similar vein a "pure money complex." See Thomas Crump, *The Phenomenon of Money* (London, 1981).

45. In *Womit handeln Banken?* Baecker speaks of an oscillation between loose and tight couplings within a network.

Stichweh, The Form of the University

1. Talcott Parsons and Gerald M. Platt, *The American University* (Cambridge, Mass., 1973).

2. Aside from the Catholic Church, the parliaments of the Isle of Man, Iceland, and Great Britain, the Monte Paschi di Siena, and approximately 70 *universities*, one can find hardly any organizations that have continuously existed

since 1520 or earlier. See Clark Kerr, "A Critical Age in the University World: Accumulated Heritage Versus Modern Imperative," *European Journal of Education* 22 (1987): 183–93. By analogy, the more than 350 years of Harvard College (founded 1636) is an unusually long period for the United States. For data, see Lubor Jílek, ed., *Historical Compendium of European Universities* (Geneva, 1984).

3. Regarding the logic of distinctions, see Francisco J. Varela, *Principles of Biological Autonomy* (New York, 1979), pp. 107–8; and Niklas Luhmann, "Wie lassen sich latente Strukturen beobachten?" in Paul Watzlawick and Peter Krieg, eds., *Das Auge des Betrachters: Beiträge zum Konstruktivismus* (Munich, 1991), pp. 61–74, especially pp. 63–65.

4. [*Fachhochschule* is a special institutional form within the German higher educational system that offers primarily vocational training (comparable to polytechnics and technical colleges) and does not award doctorates.—Trans.]

5. See Gillian Rosemary Evans, *Old Arts and New Theology: Beginnings of Theology as an Academic Discipline* (Oxford, 1980).

6. For instance, the observer might locate the overarching unity of the distinction between school and university in the fact that both are concerned with education in contrast to economic behavior or the recruitment of political loyalties.

7. See George Spencer Brown, *Laws of Form* (New York, 1972); Niklas Luhmann, "Frauen, Männer und George Spencer Brown," *Zeitschrift für Soziologie* 17 (1988): 47–71; Niklas Luhmann, Frederick D. Bunsen, and Dirk Baecker, *Unbeobachtbare Welt: Über Kunst und Architektur* (Bielefeld, 1990); Niklas Luhmann, *Die Wissenschaft der Gesellschaft* (Frankfurt a.M., 1990), especially pp. 75–87; Niklas Luhmann, "Der Gleichheitssatz als Norm und als Form," *Archiv für Rechts- und Sozialphilosophie* 77 (1991): 435–45; Niklas Luhmann, "Die Form 'Person,'" *Soziale Welt* 42 (1991): 166–75.

8. For a more detailed account, see Rudolf Stichweh, *Der frühmoderne Staat und die europäische Universität: Zur Interaktion von Politik und Erziehungssystem im Prozeß ihrer Ausdifferenzierung (16.–18. Jahrhundert)* (Frankfurt a.M., 1991), especially chap. 1.

9. See Friedrich Carl von Savigny, *System des heutigen römischen Rechts*, vol. 1 (Berlin, 1840).

10. *Akademische Gymnasien* only existed in early modern Germany. Unlike other *Gymnasien*, they not only imparted philosophical and artistic knowledge but also offered introductory courses in so-called "Faculty Studies" (e.g., law, theology). However, they were not allowed to confer degrees, nor did they enjoy "academic freedom," two characteristic privileges of universities.

11. Interesting in this context is Friedrich Carl von Savigny, *Juristische Methodenlehre* (1802/3), Jacob Grimm's edition (Stuttgart, 1951), p. 69.

12. Richard Mulcaster, *Positions Wherein Those Primitive Circumstances Be Examined, Which Are Necessarie for the Training up of Children, Either for Skill*

in Their Booke, or Health in Their Bodie (1581), ed. R. H. Quick (London, 1888), p. 252, italics mine.

13. Such a theory was forced to analyze the *individuality* of the lecturer as something that allows for a synthesis of the particular and the universal and that elicits an analogous synthesis in the addressee. See Ludwig Thilo's interesting work *Grundsätze des akademischen Vortrags, ein Beitrag zur Aufdeckung herrschender Universitätsmängel* (Frankfurt an der Oder, 1809).

14. For some preliminary considerations, see Stichweh, *Der frühmoderne Staat,* chap. 14.

15. See David Madsen, *The National University: Enduring Dream of the USA* (Detroit, 1966).

16. On self-substitution, see Niklas Luhmann, "Identitätsgebrauch in selbstsubstitutiven Ordnungen," in his *Soziologische Aufklärung 2: Aufsätze zur Theorie der Gesellschaft* (Opladen, 1975), pp. 51–71.

17. In the absence of operationally effective ideals of knowledge, this function can be framed by means of the minimal distinction between safekeeping and dilapidation (of young adults). The well-known connection between university attendance and the economic cycle substantiates the relevance of this distinction.

18. For an example of how self-evidently this reformed/unreformed distinction is employed in the periodization of English university history, see John Gascoigne, "Church and State Allied: The Failure of Parliamentary Reform of the Universities 1688–1800," in A. L. Beier et al., eds., *The First Modern Society: Essays in English History in Honour of Lawrence Stone* (Cambridge, Eng., 1989), pp. 401–29.

19. William Hamilton's well-known pronouncement reads: "All experience proves that universities, like other corporations, can only be reformed from without." William Hamilton, *Discussions on Philosophy and Literature, Education and University Reform,* 2d ed. (London, 1853), p. 448.

20. See Sheldon Rothblatt, "The Past and Future Freedom of the British University," *Minerva* 14 (1976): 251–62, especially p. 260.

21. Dieter Simon, quoted in Brigitte Mohr, "'Fachhochschulen ausbauen': Empfehlungen des Wissenschaftsrats," *Frankfurter Allgemeine Zeitung,* Dec. 19, 1990.

22. See Henrik Steffens, "Vorlesungen über die Idee der Universitäten" (1809), in Ernst Anrich, ed., *Die Idee der deutschen Universität* (Darmstadt, 1956), pp. 309–74; and John Henry Newman, *The Idea of a University* (New York, 1947).

23. Michael D. Cohen and James G. March, *Leadership and Ambiguity: The American College President* (New York, 1974), p. 195.

24. In this regard the title of the collection edited by Wilhelm Weischedel is germane: *Idee und Wirklichkeit: Dokumente zur Geschichte der Friedrich-Wilhelms-Universität zu Berlin* (Berlin, 1961).

25. Newman, *Idea of a University*, p. xxvii. Subsequently he delimits the university by reference to two features, scientific discovery and religious instruction.
26. Owen Chadwick, *Westcott and the University* (Cambridge, Eng., 1963), p. 7.
27. Just one example is Friedrich August Wolf, *Über Erziehung, Schule, Universität*, ed. Wilhelm Körte (Quedlinburg, 1835): "It is not until the university that teaching must be scientific; at schools it has to be preparatory, generally educating, and elementary" (p. 97). A detailed analysis must also take into account that university teaching should be philosophical as well (as opposed to historical, empirical, specialist), and that those two distinguishing concepts (scientific, philosophical) emphasize partially complimentary aspects of scientific cognition but also competing moments (e.g., in Schleiermacher, philosophy as distinct from particular research).
28. See Rudolf Stichweh, "Differenzierung von Schule und Universität im 18. und 19. Jahrhundert," in Stichweh, *Wissenschaft, Universität, Professionen: Soziologische Analysen* (Frankfurt a.M., 1994), pp. 193–206.
29. See Niklas Luhmann, "Die Universität als organisierte Institution," in his *Die Universität als Milieu* (Bielefeld, 1992), pp. 90–99. He adds that we are dealing here with a "specifically scientific universalism" (p. 93 n. 11), thereby stressing the combination of two "pattern variables."
30. Carl Heinrich Becker, "Vom Wesen der deutschen Universität," in Reinhold Schairer and Conrad Hoffmann, eds., *Die Universitätsideale der Kulturvölker* (Leipzig, 1925), pp. 1–30, quotation from p. 11. An excellent case study, which identifies within the development of a single scientific discipline (physics, especially mechanics) the switch to a reflexive evaluation of its own possibilities for cognition, is Kathryn M. Olesko's "The Emergence of Theoretical Physics in Germany: Franz Neumann and the Königsberg School of Physics" (Ph.D. diss., Cornell University, 1980). See also Kathryn M. Olesko, *Physics as a Calling: Discipline and Practice in the Königsberg Seminar for Physics* (Ithaca, N.Y., 1991).
31. In a similar vein, see Gerald M. Platt and Talcott Parsons, "Decision Making in the Academic System: Influence and Power Exchange," in Carlos E. Kruytbosch and Sheldon Messinger, eds., *The State of the University* (Beverly Hills, 1970), pp. 133–80, especially p. 137. For the authors, the unity of research and teaching essentially consists in forsaking the devaluation of the teaching function inherent in creating special research professorships. See also Parsons and Platt, *The American University*, p. 354.
32. For further details, see Rudolf Stichweh, "The Unity of Teaching and Research," in Stefano Poggi and Maurizio Bossi, eds., *Romanticism in Science: Science in Europe 1790–1840* (Boston, 1994), pp. 189–202.
33. For examples, see Eugene Garfield, "Research and Dedicated Mentors Nourish Science Careers at Undergraduate Institutions," *Current Contents: Social and Behavioral Sciences* 19, no. 33 (Aug. 17, 1987): 3–9.

34. See Niklas Luhmann, "Die Homogenisierung des Anfangs: Zur Ausdifferenzierung der Schulerziehung," in Niklas Luhmann and Karl Eberhard Schorr, *Zwischen Anfang und Ende: Fragen an die Pädagogik* (Frankfurt a.M., 1990), pp. 73–111; Stichweh, "Differenzierung von Schule und Universität."

35. The university's admission of women is the most distinct characteristic of the modern situation since all the other cases had late-medieval and early-modern parallels.

36. For an interesting piece of evidence, see Karl Bücher's statements regarding the establishment of and blueprints for universities in major German cities in the years prior to 1914: Karl Bücher, *Hochschulfragen: Vorträge und Aufsätze* (Leipzig, 1912); and Karl Bücher, "Neugründung von Universitäten im Deutschen Reich" (1913), in his *Die Entstehung der Volkswirtschaft* (Tübingen, 1925), pp. 469–90.

37. [In analogy to *Gesamtschule* (comprehensive school), *Gesamthochschule* is an institutional form of higher education that comprises the different traditional types of colleges—research university, professional school, *Fachhochschule*, etc.—into one conceptional, organizational, and administrative entity; it confers academic and nonacademic degrees.—Trans.]

38. In this respect England opted otherwise. With a relatively low percentage of high-school graduates entering higher education in a given year, it created a great number of new universities beginning about 1870, and consequently has by international standards small universities (fewer than 10,000 students). See Dietrich Goldschmidt, "Idealtypische Charakterisierung sieben westlicher Hochschulsysteme," *Zeitschrift für Sozialisationsforschung und Erziehungssoziologie* 11 (1991): 3–17, quotation from p. 11.

39. For a French example, see Raymond Boudon, "The French University Since 1968," *Comparative Politics* 10 (1977): 89–119, especially p. 95.

40. Friedrich Paulsen, "Überblick über die geschichtliche Entwicklung der deutschen Universitäten mit besonderer Rücksicht auf ihr Verhältnis zur Wissenschaft," in Wilhelm Lexis, ed., *Das Universitätswesen im Deutschen Reich*, vol. 1 (Berlin, 1904), pp. 3–38, quotation from p. 31. Regarding this well-calculated disregard for not so small a portion of the student body, see Rudolf Stichweh, "Akademische Freiheit, Professionalisierung der Hochschullehre und Politik," in Stichweh, *Wissenschaft, Universität, Professionen*, pp. 337–61. See also Bücher, "Die Neugründung von Universitäten im Deutschen Reich," p. 486. For Bücher, large-scale lectures at large universities have the advantage of letting weak students become aware of their own inadequacy early on.

41. According to Martin Trow, "Aspects of Diversity in American Higher Education," in Herbert J. Gans et al., eds., *On the Making of Americans: Essays in Honor of David Riesman* (Pittsburgh, 1979), pp. 271–90, quotation from p. 279.

42. These and the following figures are from Trow, "Aspects of Diversity in American Higher Education," pp. 271–72.

43. In 1991, Goldschmidt estimated the current number of American colleges

and universities to be 3,400. See his "Idealtypische Charakterisierung sieben westlicher Hochschulsysteme," p. 13.

44. I use the term in Gotthard Günther's sense. See his "Life as Poly-Contexturality" (1973), in his *Beiträge zur Grundlegung einer operationsfähigen Dialektik*, vol. 2 (Hamburg, 1979), pp. 283–306, especially p. 291.

45. Regarding the internal differentiation of the American higher-educational system, see also Gerald M. Platt, Talcott Parsons, and Rita Kirshstein, "Undergraduate Teaching Environments: Normative Orientations to Teaching Among Faculty in the Higher Educational System," *Sociological Inquiry* 48 (1978): 3–22. See as well Donald R. Light, "The Structure of the Academic Profession," *Sociology of Education* 47 (1974): 2–28, especially pp. 17–18; and Laurence R. Veysey, "Stability and Experiment in the American Undergraduate Curriculum," in Carl Daysen, ed., *Content and Context: Essays of College Education* (New York, 1973), pp. 1–63, especially pp. 5–8. Both authors stress that many ideologies of higher education that evolved in different countries at different times were institutionalized successively in the American case and still coexist today in modified forms.

46. For an overview of the emergence and diversification of research universities, see Roger L. Geiger, "The American University and Research: A Historical Perspective," in Martin Trow and Thorston Nybom, eds., *University and Society: Essays on the Social Role of Research and Higher Education* (London, 1991), pp. 200–215.

47. For a case study on imported religious diversity, see Douglas Sloan, *The Scottish Enlightenment and the American College Ideal* (New York, 1971).

48. See Paul Venabel Turner, *Campus: An American Planning Tradition* (Cambridge, Mass., 1984).

49. On the history of federal university systems, see Sheldon Rothblatt, "Historical and Comparative Remarks on the Federal Principle in Higher Education," *History of Education* 16 (1987): 151–80.

50. Numerical data are from Trow, "Aspects of Diversity in American Higher Education," p. 272.

51. For a fascinating legal and economic study of the alternatives of *financing with debt* (common in the economic system) versus *accumulation of capital* (characteristic of universities), see Henry Hansmann, "Why Do Universities Have Endowments?" *Journal of Legal Studies* 19 (1990): 3–42. That private universities accumulate capital evidently has the purpose of securing in the present the university's future reputation. This results in higher tuition for students, yet it assures them that the name of their alma mater will still denote a distinguished institution in decades to come. (See especially p. 27.)

52. For this differentiation in two-year colleges, see Steven Brint and Jerome Karable, "Institutional Origins and Transformations: The Case of American Community Colleges," in Walter W. Powell and Paul J. Dimaggio, eds., *The New Institutionalism in Organizational Analysis* (Chicago, 1991), pp. 337–60.

53. This fact is pointed out by Rothblatt, "Historical and Comparative Remarks on the Federal Principle," p. 175.

54. For England, see Rothblatt, "Historical and Comparative Remarks on the Federal Principle," pp. 164–65.

55. See Trow, "Aspects of Diversity in American Higher Education"; Martin Trow, "Elite and Mass Higher Education: American Models and European Realities," in *Research into Higher Education: Processes and Structures* (Stockholm, 1979), pp. 183–219; and Martin Trow, "The Exceptionalism of American Higher Education," in Trow and Nybom, *University and Society*, pp. 156–72.

56. On "loose coupling," see Karl E. Weick, "Educational Organizations as Loosely Coupled Systems," *Administrative Science Quarterly* 21 (1976): 1–19; J. Douglas Orton and Karl E. Weick, "Loosely Coupled Systems: A Reconceptualization," *Academy of Management Review* 15 (1990): 203–23; William Tyler, "'Loosely Coupled' Schools: A Structuralist Critique," *British Journal of Sociology of Education* 8 (1987): 313–26; William A. Firestone, "The Study of Loose Coupling: Problems, Progress, and Prospects," *Research in Sociology of Education and Socialization* 5 (1985): 3–30; and Irene S. Rubin, "Retrenchment, Loose Structure and Adaptability in the University," *Sociology of Education* 52 (1979): 211–22. On "decoupling" (of formal structure and activity structure, more about which later), see John W. Meyer and W. Richard Scott, *Organizational Environments: Ritual and Rationality* (Beverly Hills, 1983); and Martin Trow, "The Public and Private Lives of Higher Education," *Daedalus* (Winter 1975): 113–27.

57. Weick's terminology; see Karl Weick, *The Social Psychology of Organizations*, 2d. ed. (Reading, Mass., 1979); and Orton and Weick, "Loosely Coupled Systems," especially p. 205.

58. See especially the two articles by Meyer and Rowan in John W. Meyer and W. Richard Scott, *Organizational Environments: Ritual and Rationality*, updated ed. (Newbury Park, Calif., 1992), pp. 21–44, 71–97.

59. See Tyler, "'Loosely Coupled' Schools," p. 316, where he considers the theory of decoupling as an ironic reinterpretation of Max Weber's thesis of rationalization since rationality is shifted into the mythical-ceremonial realm.

60. See Fritz Heider, "Thing and Medium," *Psychological Issues* 1, no. 3 (1959): 1–34 [German 1926]; Niklas Luhmann, "Das Medium der Kunst," *Delfin* 4 (1986): 6–15.

61. For this and the following, see Stichweh, *Der frühmoderne Staat*, especially chap. 17.

62. In this context, see Veysey, "Stability and Experiment," pp. 25–26.

63. On courses and credits as a medium of university education, see also Burton R. Clark, *The Higher Education System: Academic Organization in Cross-National Perspective* (Berkeley, Calif., 1983), especially pp. 62, 105, and 281 n. 59. Clark introduces the proviso that in many national education systems such a medium exists only rudimentarily.

64. On the difference between instruction and selection theories, see Lindley

Daden and Joseph A. Cain, "Selection Type Theories," *Philosophy of Science* 56 (1989): 106–29.

65. At present these programs award about 1,000 academic degrees annually and thus far have trained perhaps 75,000 "writers." See D. G. Myers, "The Rise of Creative Writing," *Journal of the History of Ideas* 54 (1993): 277–97.

66. Parsons and Platt, *The American University*.

67. Talcott Parsons, "Introduction to Part Two: Differentiation and Variation in Social Structures," in Talcott Parsons et al., eds., *Theories of Societies* (New York, 1961), pp. 239–64, quotation from p. 261, italics mine.

68. See Max Weber, *The Protestant Ethic and the Spirit of Capitalism*, trans. Talcott Parsons (London, 1930), pp. 13–31; Weber only briefly mentions institutions of higher learning ("rational, systematic, and specialized pursuit of science"), on p. 14.

69. See also Rudolf Stichweh, "Rationalität bei Parsons," *Zeitschrift für Soziologie* 9 (1980): 54–78.

70. See, for instance, Herbert A. Simon, "Invariants of Human Behavior," *Annual Review of Psychology* 41 (1990): 1–19.

71. As an example of this logic of extending what can be included in the concept of rationality, see Hilary Putnam, "The Impact of Science on Modern Conceptions of Rationality," *Synthese* 46 (1981): 359–82. He explains the success of modern science on the basis of the informal rationality of its methods (especially pp. 377–78).

72. Using Daniel Dennett's theory as an example, Stephen P. Stich elaborates on this in "Could Man Be an Irrational Animal? Some Notes on the Epistemology of Rationality," *Synthese* 64 (1985): 115–35.

73. See, for instance, the following statement by the Prussian minister of cultural affairs, Carl Heinrich Becker, conspicuously formulated under Max Weber's influence: "Only he who remains aware of the deep tension between the entirely and thoroughly rationalized methods and practical purposes of the German university and its completely irrational, ideal, and unitary final goal can understand its essence." Becker, "Vom Wesen der deutschen Universität," p. 2.

Willke, The Contingency and Necessity of the State

1. See Iring Fetscher, "Einleitung," in Thomas Hobbes, *Leviathan* (Frankfurt, 1984), p. xxxv.

2. This also could be demonstrated for the third side of the form, that is to say, the totalitarian integration of people through merciless education.

3. Georg W. F. Hegel, *Elements of the Philosophy of Right*, trans. H. B. Nisbet (Cambridge, Eng., 1991), p. 220 (§182).

4. Ibid.

5. Ibid.

6. George Spencer-Brown, *Laws of Form* (New York, 1979), p. 1.

7. In Spencer-Brown's sense; see ibid., pp. 69ff., 102ff.

8. J. P. Nettle, "The State as a Conceptual Variable," *World Politics* 20 (1968): 559–94, quotation from p. 562.

9. For an unsurpassed account of the mythology of the state, see Ernst Cassirer, *The Myth of the State* (New Haven, Conn., 1946). Theda Skocpol has characterized the new, "Tocquevillian" view succinctly: "When the effects of states are explored from the Tocquevillian point of view, those effects are *not* traced by dissecting state strategies or policies and their possibilities for implementation. Instead, the investigator looks more macroscopically at the ways in which the structures and activities of states unintentionally influence the formation of groups and the political capacities, ideas, and demands of various sectors of society." Theda Skocpol, "Bringing the State Back In," in Peter Evans et al., eds., *Bringing the State Back In* (Cambridge, 1985), pp. 19–37, quotation from p. 21. However, the question remains whether certain structures and activities that are attributed to the state can effectuate *unintentional* changes in social conditions and problem patterns.

10. "The modern state and its supporting code of power and law can therefore be conceived as a way of differentiating and cooling down the special paradoxes of self-reference." Niklas Luhmann, *Soziologische Aufklärung 4: Beiträge zur funktionalen Differenzierung der Gesellschaft* (Opladen, 1987), p. 163.

11. Ernst Forsthoff, *Der Staat der Industriegesellschaft* (Munich, 1971), especially pp. 82ff.

12. Incidentally, this is contrary to Hegel's explicit formulations; see especially Hegel, *Elements of the Philosophy of Right*, pp. 282–83 (§260).

13. [*Stasi*: abbreviation for "Staatssicherheitsdienst," the GDR's state security organization.—Trans.]

14. Eugen Kogon, *The Theory and Practice of Hell: The German Concentration Camps and the System Behind Them*, trans. Heinz Norden (New York, 1950). For an approach based on cynicism, see Peter Sloterdijk, *Critique of Cynical Reason*, trans. Michael Eldred (Minneapolis, 1987), pp. 241–43.

15. Sloterdijk, *Critique of Cynical Reason*, p. 245, translation amended.

16. Detlef Pollack provides astute thoughts on this issue in "Das Ende einer Organisationsgesellschaft: Systemtheoretische Überlegungen zum gesellschaftlichen Umbruch in der DDR," *Zeitschrift für Soziologie* 19 (1990): 292–307. Regarding the health-care system of the GDR and some of the effects of the state-induced "infantilization," see Josef Düllings, "Staatsdominanz und ihre Folgen für die Entwicklung der ambulanten Versorgung Ostdeutschlands," *Medizin, Mensch, Gesellschaft* 16 (1991): 13–20.

17. [SED: abbreviation for *Sozialistische Einheitspartei Deutschlands* (United Socialist Party of Germany); the ruling and *de facto* state party of the GDR.—Trans.]

18. For a case study, see Helmut Willke, *Leitungswissenschaft in der DDR* (Berlin, 1979). This also serves as counterevidence to the frequent argument that

scientific research, as usual, only identifies the process of "inevitable" perversion after the fact.

19. Jürgen Habermas's comment on the Marxist theory of revolution is apropos: "But if emancipation and reconciliation are represented only in the mode of a *de-differentiation* of hypercomplex conditions of life, it is quite easy for systems theory, in view of stubborn complexities, to dismiss the unifying power of reason as a sheer illusion." Jürgen Habermas, *The Philosophical Discourse of Modernity: Twelve Lectures*, trans. Frederick G. Lawrence (Cambridge, Mass., 1987), p. 67. Stephen Holmes encapsulates this insight in the catchy phrase "No democracy without differentiation." Stephen Holmes, "Poesie der Indifferenz," in Dirk Baecker et al., eds., *Theorie als Passion: Niklas Luhmann zum 60. Geburtstag* (Frankfurt a.M., 1987), pp. 15–45, quotation from p. 25.

20. For more detail, see Helmut Willke, *Ironie des Staates* (Frankfurt a.M., 1992), especially chap. 2.

21. Richard Rorty, *Contingency, Irony, and Solidarity* (Cambridge, Eng., 1989).

22. "I know that one day people will live, for whom the problems that cause us the most anguish today will no longer even exist. This is my fate, which I have to assume and which I do assume. But this cannot reduce me to despair or to catatonic ruminations." Cornelius Castoriadis, *The Imaginary Institution of Society*, trans. Kathleen Blamey (Cambridge, Mass., 1987), 92–93.

23. Rorty, *Contingency, Irony, and Solidarity*, p. xv. A detailed definition is given on p. 73: "I shall define an 'ironist' as someone who fulfills three conditions: (1) She has radical and continuing doubts about the final vocabulary she currently uses, because she has been impressed by other vocabularies, vocabularies taken as final by people or books she has encountered; (2) she realizes that argument phrased in her present vocabulary can neither underwrite nor dissolve these doubts; (3) insofar as she philosophizes about her situation, she does not think that her vocabulary is closer to reality than others, that it is in touch with a power not herself."

24. Ibid., p. 92.

25. See ibid., p. 59.

26. This leads Niklas Luhmann to the conclusion "that freedom is only possible as freedom to evil since everything that is good corresponds to God's will." Niklas Luhmann, "Sthenographie und Euryalistik," in Hans Ulrich Gumbrecht and K. Ludwig Pfeiffer, eds., *Paradoxien, Dissonanzen, Zusammenbrüche: Situationen offener Epistemologie* (Frankfurt a.M., 1991), pp. 58–82, quotation from p. 64. Or to cite a formulation by Johanne Villeneuve: "The devil causes the category of the possible, time and again, to impact on the unmistakable strength of the divine law." Johanne Villeneuve, "Der Teufel ist ein Spieler oder: Wie kommt ein Eisbär an die Adria?" in Gumbrecht and Pfeiffer, *Paradoxien, Dissonanzen, Zusammenbrüche*, pp. 83–95, quotation from p. 85.

27. Søren Kierkegaard, *The Concept of Irony with Continual Reference to*

Socrates, ed. and trans. Howard V. Hong and Edna H. Hong (Princeton, N.J., 1992), p. 6 (thesis 15).

Japp, The Form of Protest in the New Social Movements

I am grateful to Dirk Baecker for several suggestions on this topic which I have utilized freely and without indicating their authorship.

1. See K. P. Japp, "Selbsterzeugung oder Fremdverschulden? Thesen zum Rationalismus in den Theorien sozialer Bewegungen," *Soziale Welt* 35 (1984): 313–29; and H. W. Ahlemeyer, "Was ist eine soziale Bewegung?" *Zeitschrift für Soziologie* 18 (1989): 175–91.

2. [Here and passim, *Betroffenheit* has been rendered as "concern" in the sense of "feeling" or "being concerned." While "concern" better fits the topic under discussion than standard translations such as "dismay," "consternation," or "disconcertedness," it still does not fully convey the cultural meaning and function *Betroffenheit* has taken on in West German public discourse since the 1970s. Given the context of the new social movements, *Betroffenheit* denotes on a psychological level an emotive response to the risky nature of modern societies, most notably their technological potential for self-destruction. *Betroffenheit* counteracts the anonymous and "invisible" reality of those risks and threats by bringing them back into the emotional and pragmatic realm of the individual: On one hand, *Betroffenheit* makes the individual feel negatively affected by (as if he were the victim of) an event (say, a nuclear meltdown), although the event has not yet "hit" him personally. On the other hand, whatever makes the individual feel *betroffen* (concerned) is the product of human agency; that is to say, there are people to be held responsible for it. Hence, earthquakes or imaginary invasions by Martians are not suitable causes of *Betroffenheit*. Apart from signifying a state of mind—a mild form of anxiety—*Betroffenheit* constitutes a particular form of communication. Feeling concerned should not be kept to oneself, but calls for being shared with and by others. Moreover, it acquires the moral status of virtue and imperative: you ought to be concerned! In this aspect, the discourse of *Betroffenheit* continues the eighteenth-century discourse of "compassion," "empathy," and "sensitivity." Communication of *Betroffenheit*, as Klaus P. Japp points out, can motivate political involvement and serve as the rationale for protest movements. See, for example, Monika Sperr's political biography, *Petra Karin Kelly: Politikerin aus Betroffenheit* (Munich, 1983); and Karlheinz Kress and Klaus Günther Nikolai, *Bürgerinitiativen: Zum Verhältnis von Betroffenheit und politischer Beteiligung der Bürger* (Bonn, 1985). Communication of *Betroffenheit*, however, need not lead to political engagement. It can turn into a self-perpetuating discourse in which talking about one's *Betroffenheit* produces primarily more talk about *Betroffenheit*, irrespective of the legitimacy and consequences of one's own concern. Paradoxically, communicating this unhappy state of *Betroffenheit* allows the individual or group to feel good about themselves. On this

ritualistic, even faddish aspect of *Betroffenheit*, see Cora Stephan, *Der Betroffenheitskult: Eine politische Sittengeschichte* (Berlin, 1993).—Trans.]

3. Niklas Luhmann, *Soziologie des Risikos* (Berlin, 1991).

4. The unity of the distinction thus exists in the mode of the reflexive self-reference of society. This fact remains latent for the new social movements insofar as they observe "Society" and cannot (and must not) also observe this distinction that enables their observing.

5. See George Spencer-Brown, *Laws of Form* (reprint, New York, 1979).

6. Society *fades*, as it were, into the issue against which the protest is directed.

7. See Luhmann, *Soziologie des Risikos*. A similar though differently reasoned argument is put forth by authors who employ the term "latent networks." See, for instance, A. Melucci, *Nomads of the Present* (London, 1989).

8. And how could one possibly explain to anyone, in keeping with political sincerity, that one only needs the issue so that one is able to protest?

9. To speak of effective blocking instead of strict conditioning would not necessarily be inaccurate, but the former expression too-readily evokes associations with, for example, "fundamentalism" or "sectarianism." See the references below in note 19.

10. The new social movements are first-order observers. They observe the world from the viewpoint of an innocent realist; Mary Douglas refers to this as "innocent culture." See Mary Douglas, "Risk as a Forensic Resource: From 'Chance' to 'Danger,'" *Daedalus* 119 (1990): 1–16.

11. Or again: the resistance to latency.

12. See Luhmann, *Soziologie des Risikos*.

13. Ibid., chap. 12.

14. To which I limit myself here.

15. By "sides" I refer here to acceptable/unacceptable, safe/unsafe, responsible/irresponsible, etc.

16. This problem is formally analogous to the one of establishing rationality of action in formal organizations. See K. P. Japp, "Selbstverstärkungseffekte riskanter Entscheidungen: Zur Unterscheidung von Rationalität und Risiko," *Zeitschrift für Soziologie* 21 (1992): 31–48.

17. K. P. Japp and Jost Halfmann, "The New Social Movements as Active Observers of Risk," *Social Science Information* 32, no. 3 (1993): 427–46.

18. Such repetitions of consent could eventually become boring. R. Paris suggests that this constitutes an important reason for the periodic collapse of activity of new social movements. See R. Paris, "Situative Bewegung: Moderne Protestmentalität und politisches Engagement," *Leviathan* 3 (1989): 322–36.

19. In *Risk and Culture* (Berkeley, Calif., 1983), M. Douglas and A. Wildavsky speak of "sects."

20. For a more detailed account, see K. P. Japp, "Risiken der Technisierung und die neuen sozialen Bewegungen," in G. Bechmann, ed., *Risiko und Gesellschaft* (Opladen, 1992), pp. 375–402.

21. See G. Teubner, "Hyperzyklus in Recht und Organisation: Zum Verhältnis von Selbstbeobachtung, Selbstkonstitution und Autopoiesis," in H. Haferkamp and M. Schmidt, eds., *Sinn, Kommunikation und soziale Differenzierung* (Frankfurt a.M., 1987), pp. 89–128.

22. Charles Perrow, *Normal Accidents: Living with High-Risk Technologies* (New York, 1984), pp. 325–28.

23. See Paris, "Situative Bewegung," p. 326.

24. See A. O. Hirschman, *Shifting Involvements* (Princeton, N.J., 1982).

25. In the meantime, however, worry seems to prevail. See, for example, B. Wynne, "Unruly Technology: Practical Rules, Impractical Discourses and Public Understanding," *Social Studies of Science* 18 (1988): 147–67.

26. P. Fuchs, *Erreichbarkeit der Gesellschaft* (Frankfurt a.M., 1992), pp. 101.

27. Douglas and Wildavsky (see *Risk and Culture*) therefore distinguish between this distinction (society/environment) and the one between center and periphery. This by no means implies, however, that each function system cannot observe "its" specific difference from the environment of society. See Niklas Luhmann, *Ecological Communication*, trans. John Bednarz, Jr. (Chicago, 1989). These observations could then be condensed with those of other function systems into one observational context.

28. Fuchs, *Erreichbarkeit der Gesellschaft*, p. 237.

29. Members of the new social movements do not run for office but leave argumentative conflicts to the political parties.

30. See Jay D. Starling, *Municipal Coping Strategies: "As Soon as the Dust Settles"* (Beverly Hills, Calif., 1985).

31. Niklas Luhmann, *Die Wissenschaft der Gesellschaft* (Frankfurt a.M., 1990), p. 79.

32. See, for instance, B. Wynne, "Institutional Mythologies and Dual Societies in the Management of Risk," in Howard C. Kunreuther and Eryl V. Ley, eds., *The Risk Analysis Controversy: An Institutional Perspective* (Berlin, 1982).

33. I take the term "coloring" (*Kolorierung*) and its meaning from Niklas Luhmann, "Wie lassen sich latente Strukturen beobachten?" in Paul Watzlawick and Peter Krieg, eds., *Das Auge des Betrachters: Beiträge zum Konstruktivismus* (Munich, 1991), pp. 61–74.

34. See Fuchs, *Erreichbarkeit der Gesellschaft*.

35. See Luhmann, *Soziologie des Risikos*.

36. See Luhmann, "Wie lassen sich latente Strukturen beobachten?"

37. The movements thereby provide a way to attribute all problems to "Society" and all solutions to themselves. This is another safety device protecting against the possibility of perceiving society's side as more than an endangerment, however catastrophic it may be. And in this regard, that is, in avoiding second-order observations of solutions that are accompanied by problems, the new social movements have an affinity with observational forms of the socialist labor movement.

38. See Fuchs, *Erreichbarkeit der Gesellschaft*.

39. A sort of permanent all points bulletin is put out for culprits; the more intense the search becomes, the more pointless it actually is.

40. Japp, "Risiken der Technisierung."

41. K. P. Japp, "Kollektive Akteure als soziale Systeme?" in H.-J. Unverferth, ed., *System und Selbstproduktion* (Frankfurt a.M., 1986), pp. 166–91.

42. Compare, for example, the notion of the "practically negligible residual risk" in the legal communication medium: U. K. Preuß, "Sicherheit durch Recht: Rationalitätsgrenzen eines Konzepts," *Kritische Vierteljahresschrift für Rechtswissenschaft und Gesetzgebung* 4 (1989): 3–26.

43. See Niklas Luhmann, *Social Systems*, trans. John Bednarz, Jr., with Dirk Baecker (Stanford, Calif., 1995), pp. 357–404.

44. In this respect, the women's movement has a much harder lot. The probability that their "disasters" will occur is extremely high; women are directly faced with them in their daily lives. This leads to differentiating and generalizing issues that might provide different possibilities for internal reflection.

45. See Luhmann, *Soziologie des Risikos*.

46. K. P. Japp, "Neue sozial Bewegungen und die Kontinuität der Moderne," in J. Berger, ed., *Die Moderne: Kontinuitäten und Zäsuren* (Göttingen, 1986), pp. 311–34. See also Hirschman's *Shifting Involvements*, which emphasizes several times that the objective in the shift from the individualistic enhancement of options to a collectivity based on solidarity (thus unity) lies primarily in collectivity per se and only secondarily in the collective goals sought. That makes the new social movements a value-rational community. A commentary in the *Frankfurter Allgemeine Zeitung* in the summer of 1992 even suggested that the eruptions of violence in Rostock signaled a search for restricting one's own behavior through state power.

47. See Fuchs, *Erreichbarkeit der Gesellschaft*, and Japp, "Risiken der Technisierung."

48. Thus it could be asked: How is it possible *not* to observe latency?

49. The new social movements hence are not capable of rationality if we take rationality to mean the ability to productively reinternalize environmental effects into operations (see Luhmann, *Social Systems*). On the question of whether a *different* rationality can be attributed to these movements, that is, a rationality related to their specific systemic contexts, see Japp, "Risiken der Technisierung."

50. Take the case of civil rights movements in the former German Democratic Republic.

51. See Melucci, *Nomads of the Present*.

52. See Douglas and Wildavsky, *Risk and Culture*. I place the focus of risk perception within the domain of central social institutions (such as the state and the market), since the issue is society's observation.

53. See, for example, J. C. Jenkins, "Resource Mobilization Theory and the Study of Social Movements," *Annual Review of Sociology* 9 (1983): 527–53.

54. See Ahlemeyer, "Was ist eine soziale Bewegung?" and Japp, "Kollektive Akteure als soziale Systeme?"

55. Among other things, I refer here to blockades of waste-disposal facilities that are considered dangerous. For the boundless literature on "facility siting," that is, the choice of sites for facilities perceived as dangerous, see F. N. Laird, "The Decline of Deference: The Political Context of Risk Communication," *Risk Analysis* 9, no. 4 (1989): 543–50.

56. See Laird, "Decline of Deference"; S. G. Hadden, "Public Perception of Hazardous Waste," *Risk Analysis* 11, no. 1 (1991): 47–57; and O. Renn and D. Levine, "Trust and Credibility in Risk Communication," in R. E. Kasperson and P. M. Wiedemann, eds., *Themes and Tasks of Risk Communication* (Jülich, 1988), pp. 51–81.

57. Wynne ("Institutional Mythologies and Dual Societies"), for example, argues for techniques that attempt to increase "institutional credibility."

58. See Hadden, "Public Perception of Hazardous Waste."

59. See Laird, "Decline of Deference."

60. See Niklas Luhmann, *Observations on Modernity*, trans. William Whobrey (Stanford, Calif., 1998).

61. What comes to mind in the first place are distinctions such as the one between ecology and economy in the economic system, the one between ecology and politics in the political system, or the one between ecology and science in the scientific system. All those distinctions can "work themselves out" by means of—among other factors—perturbations from the new social movements.

62. Max Miller, "Rationaler Dissens: Zur gesellschaftlichen Funktion sozialer Konflikte," H.-J. Giegel, ed., *Kommunikation und Konsens in modernen Gesellschaften* (Frankfurt a.M., 1992), pp. 31–58.

63. See Luhmann, *Observations on Modernity*.

64. A. Hahn, "Verständigung als Strategie," in Max Haller, H.-J. Hoffmann-Nowotny, and Wolfgang Zapf, eds., *Kultur und Gesellschaft: Verhandlungen des 24. Deutschen Soziologentags, des 11. Östereichischen Soziologentags und des 8. Kongress der Schweizerischen Gesellschaft für Soziologie in Zürich 1988* (Frankfurt a.M., 1989), pp. 346–59.

Corsi, The Dark Side of a Career

1. See Alois Hahn, "Zur Soziologie der Beichte und anderer Formen institutionalisierter Bekenntnisse: Selbstthematisierung und Zivilisationsprozeß," *Kölner Zeitschrift für Soziologie und Sozialpsychologie* 34 (1982): 408–34. In modernity this tendency became still more pronounced. As an example, see Jacques Le Brun, "Conversion et continuité intérieure dans les biographies spirituelles française du XVIIème siècle," in *La conversion au XVIIe siècle: Actes du XIIe colloque de Marseille du Centre Méridional de Rencontres sur le XVIIe siècle* (Marseille, 1983): "The time of the conversion permits affirmation of the self,

whereas continuities lead to the disintegration of the individual within the social network, of psychological constancy, or of the providential plan. . . . In a conversion it is an 'I' that forces the decision" (p. 327).

2. On stratification, see Niklas Luhmann and Raffaele De Giorgi, *Teoria della società* (Milan, 1992), pp. 281ff.

3. See Jacques Le Goff, "Temps de l'Eglise et temps du marchand," *Annales* 15 (1960): 417–33.

4. See Jan Romein, *Die Biographie: Einführung in ihre Geschichte und ihre Problematik* (Bern, 1948), p. 28. According to Romein, the biography genre experienced a series of particularly favorable moments: In the fifth century B.C., following the invention of the phonetic script, the Greeks (e.g., Critias, Xenophon, Plutarch) began to consider a chronology of life relevant for many areas. Subsequent important moments are the emergence of the Roman Empire in the second century A.D. and the emergence of modernity in the sixteenth century. The sociological significance of these epochal thresholds may suggest that the evolution of the biography was not a chance event. See also Albrecht Dihle, *Studien zur griechischen Biographie* (Göttingen, 1970). Dihle views Socrates' apology as the first attempt to interpret the person in reference to his or her life. The biography then was regarded, perhaps for the first time, as an example of virtue.

5. The word "career," at least in its present meaning, was coined in seventeenth-century France, spread to Italy in the beginning of the eighteenth century, and later reached the rest of Europe.

6. See James S. Coleman et al., *Equality of Educational Opportunity* (Washington, D.C., 1966). But what does "equality of educational opportunity" mean? Should the school be made the same for all (the pupils will differentiate themselves), or should the pupils be considered as on a par with each other (and the schools to be different)?

7. "Professional life appears as a series of thresholds, of stages, and bifurcations whose course is indicated by career; or better yet, career is the very sequence of titles, roles, and distinctions to the extent that occupation (and not personal talent, family, chance, or other circumstances) determines its chronology." Jean-René Treanton, "Le concept de carrière," *Revue Française de Sociologie* 1 (1960): 73–80, quotation from p. 73. The definition of "career" I employ here is based on the following study: Niklas Luhmann and Karl-Eberhard Schorr, *Reflexionsprobleme im Erziehungswesen* (Frankfurt a.M., 1982), pp. 277ff. See also Jeff Hearn, "Toward a Concept of Non-Career," *Sociological Review* 25 (1977): 273–88. Hearn defines "pure career" as follows: "The pure career is a structuring of time in the past and in the future. It is concerned with justifications, explanations and certain knowledge in the past; and with expectations, anticipations and uncertainties in the future" (p. 276).

8. See Eviatar Zerubavel, "Private Time and Public Time: The Temporal Structure of Social Accessibility and Professional Commitments," *Social Forces* 58 (1979): 38–58. According to Zerubavel, in modern societies the difference be-

tween private and public is temporalized and transformed into the difference between leisure and work. In this view, the function of an occupation is to make individuals publicly observable. Zerubavel also notes that certain careers negate "one's right to be inaccessible at certain times" (p. 43). This is the case in hospitals and prisons, which are not accidentally described as total institutions. The difference between private and public signifies different weights attributed to self-selection and external selection. The career entails both while leisure consists almost exclusively in self-selection.

9. The number of excluded possibilities depends on the complexity the career has already realized.

10. See Niklas Luhmann, *Social Systems*, trans. John Bednarz, Jr., with Dirk Baecker (Stanford, Calif., 1995), pp. 351–52.

11. With "re-entry" I refer to George Spencer Brown, *Laws of Form* (London, 1969). See also Elena Esposito, *L'operazione di osservazione: Construttivismo e teoria dei sistemi social* (Milan, 1992).

12. Niklas Luhmann and Karl-Eberhard Schorr use the term "selection medium" to indicate that the medium triggers the production of single selections and thereby the creation of careers. See Luhmann and Schorr, *Reflexionsprobleme im Erziehungswesen*, pp. 300ff. On the notion of medium, see Fritz Heider, "Thing and Medium," *Psychological Issues* 1, no. 3 (1959): 1–34 [German 1926]. A medium consists of a number of loosely coupled elements that, when tightly coupled, lead to the generation of forms. For my purposes the selection possibilities can be called a medium in which forms are only imprinted if selections are related to social positions.

13. Even in the domain of leisure, one encounters such situations: decisions on alternative vacation plans are frequently made on the basis of how others make theirs.

14. "Every instance of binding time has a social price." Niklas Luhmann, *Soziologie des Risikos* (Berlin, 1991), p. 66.

15. See Spencer Brown, *Laws of Form*.

16. Luhmann and De Giorgi, *Teoria della società*, p. 17. See also Esposito, *L'operazione di osservazione*.

17. That this formulation is paradoxical—defining *identity* by what it is not—is not important at the moment. In any case, see Niklas Luhmann, "Identiät: Was oder wie?" in his *Soziologische Aufklärung 5: Konstruktivistische Perspektiven* (Opladen, 1990), pp. 14–30.

18. A landowner can go on a hunt or look after his pension, yet the question remains as to what social relevance he acquires in the process.

19. I owe this suggestion to Niklas Luhmann.

20. The potential significance of the age connection for our society becomes noticeable when modern societies are compared to others. In ancient Rome, for instance, becoming autonomous from one's birth family did not occur at a certain age or upon personal achievements in battle or with the attainment of other mer-

its. Only when a son formed a new family was he no longer called an *adolescens* and did his father lose rights over him. See Paul Veyne, *The Roman Empire*, in Philippe Ariès and Georges Duby, eds., *A History of Private Life. Vol. I: From Pagan Rome to Byzantium*, trans. Arthur Goldhammer (Cambridge, Mass. 1987), pp. 5–234. The conception of life as a sequence of phases and corresponding expectations is an invention of modernity. On the discovery of the child, see Philippe Ariès, *Centuries of Childhood: A Social History of Family Life*, trans. Robert Baldick (London, 1962). On the discovery of the difference between childhood, youth, and so forth see Tamara K. Hareven, "The Life Course and Aging in Historical Perspective," in T. K. Hareven and Kathleen J. Adams, eds., *Aging and Life Course Transitions: An Interdisciplinary Perspective* (London, 1982), pp. 1–53.

21. See Burton R. Clark, "The 'Cooling-Out' Function of Higher Education," *American Journal of Sociology* 65 (1960): 569–76. On the relationships between age and job satisfaction, see Susan R. Rhodes, "Age-Related Differences in Work Attitudes and Behavior: A Review and Conceptual Analysis," *Psychological Bulletin* 93 (1983): 328–67. Rhodes notes a "positive relationship between age and overall job satisfaction, satisfaction with work itself, and job involvement." In addition, she mentions that constellations of fault attribution also vary with age: "Whereas inexperience and perhaps lack of caution seem to be associated with accidents among younger workers, decline in psychological functioning, such as sensorimotor response, is more salient for older workers" (p. 355).

22. With regard to the notion of medium, we could say that age is the condition for the fluidity of the medium and that aging causes a sort of rigidity of the internal connections such that imprintings of new forms become more difficult.

23. See Niklas Luhmann, "'Distinctions directrices': Über Codierung von Semantiken und Systemen," in his *Soziologische Aufklärung 4: Beiträge zur funktionalen Differenzierung der Gesellschaft* (Opladen, 1987), pp. 13–31.

24. If the distinction between career and age is collapsed onto the side of age, it appears to retransform into the one between time and eternity, in which ultimately everything becomes futile (pointless).

25. See A. I. Rabin, "Future Time Perspective and Ego Strength," in J. T. Frazer, N. Lawrence, and D. Park, *The Study of Time: Proceedings of the Conference of the International Society for the Study of Time*, vol. 3 (Berlin, 1978), pp. 294–306. Rabin stresses the positive relationships between perspectives of the future and tendencies to delay gratification, as well as those between the ability to orient oneself in time and the ability to keep responsibility under one's personal control. The lack of such perspectives and orientations, however, is associated with impulsivity, meaning that the projection of a future is a mandatory condition for the improbability of human action.

26. This especially pertains to standardized career patterns, which, however, function less and less as models. The strong homogenization based on age can also lead to problematic situations, such as the concentration of the professional

life-phase in the middle years, which in an extreme case can lead to poverty or to the bankruptcy of the welfare state. See Fred Best, "The Time of Our Lives: The Parameters of Lifetime Distribution of Education, Work, and Leisure," *Loisir et Société* 1 (1978): 95–124. On the sequences of events in careers that are considered "normal" and for which a certain conformity is expected, see Dennis P. Hogan, "The Variable Order of Events in the Life Course," *American Sociological Review* 43 (1978): 573–86.

Simon, The Other Side of Illness

1. George Spencer-Brown, *Laws of Form* (reprint, New York, 1979). In this context, see also Fritz B. Simon, *Unterschiede, die Unterschiede machen. Klinische Epistemologie: Grundlage einer systemischen Psychatrie und Psychosomatik* (Heidelberg, 1988).

2. See Humberto R. Maturana, *Explanation and Reality* (Heidelberg, 1992) (lecture on audiotape).

3. Unless otherwise indicated, my presentation of the historical development follows Wolfgang Eckart, *Geschichte der Medizin* (Heidelberg, 1990).

4. From the Greek *sympiptein*: to coincide, to occur simultaneously.

5. Henceforth, I will italicize certain signifiers in order to distinguish them from the phenomena designated. Thus, *patient* is the signifier for a human being considered as suffering (see Latin *patiens*: suffering) and not the human being himself. It is then necessary to ask what the characteristics are to which he owes this designation and what the consequences of this signification are for interactions and communication.

6. If one follows Spencer-Brown and distinguishes between (1) distinctions within a first phenomenal domain and (2) distinctions as signs of those first distinctions, then symptoms are to be interpreted as signs of another distinction within a different phenomenal domain.

7. See Eckart, *Geschichte der Medizin*, p. 8.

8. Claudius Galenos of Pergamum (A.D. 129–99).

9. Morgagni, quoted in Herwig Hamperl, *Lehrbuch der allgemeinen Pathologie and pathologischen Anatomie*, 28th ed. (Berlin, 1968), p. 1.

10. See Michel Foucault, *The Birth of the Clinic: An Archaeology of Medical Perception*, trans. A. M. Sheridan Smith (New York, 1973).

11. See Niklas Luhmann, *Social Systems*, trans. John Bednarz, Jr., with Dirk Baecker (Stanford, Calif., 1995).

12. See Karl Jaspers, *General Psychopathology*, trans. J. Hoenig and Marian W. Hamilton (Chicago, 1963), 302–3. See also his criticism of psychoanalysis, which, according to Jaspers, does not respect the boundary between understanding and explaining: "Making signification absolute and reducing signification to one single level of understanding meaning amounts to a 'worldview' for which everything turns into but one kind of symbol, namely one that can be interpreted.

Interpretation is extended from factual hysterical symptoms and other graspable indications of disorder to all diseases whatsoever, to the entire human biography." Karl Jaspers, "Zur Kritik der Psychoanalyse," *Nervenarzt* 21 (1950): 465–68.

13. Fifth through twelfth centuries.

14. See Claude Lévi-Strauss, *Structural Anthropology*, trans. Claire Jacobson and Brooke Grundfest Schoepf (New York, 1963), pp. 197–98.

15. From the Greek *aitia*: responsibility; guilt, blame, accusation.

16. See Fritz B. Simon, *My Psychosis, My Bicycle, and I*, trans. S. and B. Hofmeister (Northvale, N.J., 1996).

17. See Humberto R. Maturana, "The Organization of the Living: A Theory of the Living Organization," *International Journal of Man-Machine Studies* 7 (1975): 313–32.

18. See especially Ludwik Fleck, *Genesis and Development of a Scientific Fact* (1935), trans. Frederick Bradley and Thaddeus J. Trenn (Chicago, 1979), pp. 21–22. Fleck shows how collective processes lead to the assumption of such units, to which the cause of symptoms are attributed. Foucault characterizes the classificatory creation of disease entries in terms of a "botanical model"; see Foucault, *Birth of the Clinic*, pp. 7–9.

19. "Punctuation refers to the structuring and organization by an observer of a continuous sequence of events and behaviors. . . . The manner in which an ongoing communication process and/or interaction sequence is punctuated determines the meaning attributed to it and how each person's behavior will be evaluated, that is, who is responsible or 'guilty,' and how one decides to (re)act." Fritz B. Simon, Helm Stierlin, and Lyman C. Wynne, *The Language of Family Therapy: A Systematic Vocabulary and Sourcebook* (New York, 1985), p. 284.

20. See Michel Foucault, *Mental Illness and Psychology*, trans. Alan Sheridan (New York, 1976).

21. Ibid., p. 67, translation slightly modified.

22. Ibid., p. 68.

23. See ibid., p. 71.

24. F. Kluge, *Etymologisches Wörterbuch der deutschen Sprache* (1883), 21st ed. (Berlin, 1975).

25. Erving Goffman defines "psychotic behavior" from a sociological stance as "a failure to abide by rules established for the conduct of face-to-face interaction." See Erving Goffman, *Interaction Ritual: Essays in Face-to-Face Behavior* (Chicago, 1967), p. 141.

26. According to Wilhelm Griesinger (1845), quoted from Eckart, *Geschichte der Medizin*, p. 239.

27. Ludwik Fleck shows how, by means of the "Wassermann reaction," the explanation of syphilis as the result of a spirochetal infection acquired the status of a "fact" and how certain research methods lead to atomistic descriptions. See Fleck, *Genesis and Development of a Scientific Fact*, pp. 52–81.

28. See Thomas Stephen Szasz, *The Myth of Mental Illness: Foundations of a Theory of Personal Conduct* (New York, 1961); and Thomas Stephen Szasz, *The Myth of Psychotherapy: Mental Healing as Religion, Rhetoric, and Repression* (Garden City, N.Y., 1978).

29. On this point, see Fritz B. Simon, "Das deterministische Chaos schizophrenen Denkens," *Familiendynamik* 14 (1989): 236–58; and Simon, *My Psychosis, My Bicycle, and I*.

30. There is by now a rich literature on systemic therapy; for an overview, see Lynn Hoffman, *Foundations of Family Therapy* (New York, 1981); Simon, Stierlin, and Wynne, *Language of Family Therapy*; and Helm Stierlin, Fritz B. Simon, and Günther Schmidt, eds., *Familiar Realities* (New York, 1987).

31. See Humberto R. Maturana, "Repräsentation und Kommunikation," in his *Erkennen: Die Organisation und Verkörperung von Wirklichkeit*, trans. Wolfgang K. Köck (Braunschweig, 1982), pp. 272–96.

32. See Francisco J. Varela, *Principles of Biological Autonomy* (New York, 1979), pp. 211–37.

33. See Heinz von Foerster, "Abbau und Aufbau," in Fritz. B. Simon, ed., *Lebende Systeme* (Berlin, 1988), pp. 19–33; and Heinz von Foerster, "Objects: Token for (*Eigen-*)Behaviors," in his *Observing Systems* (Seaside, Calif., 1981), pp. 273–85.

34. See F. B. Simon, *Unterschiede, die Unterschiede machen*, pp. 61–62, 100.

Library of Congress Cataloging-in-Publication Data

Problems of form / edited by Dirk Baecker : translated by Michael Irmscher, with Leah Edwards.
     p.   cm. — (Writing science)
  ISBN 0-8047-3423-2 (cloth : alk. paper). — ISBN 0-8047-3424-0 (pbk. : alk. paper)
    1. Social systems.   2. System theory   3. Mathematical sociology.   4. Spencer-Brown, G.   Laws of form.   5. Logic, Symbolic and mathematical.   6. Form (Logic)   I. Baecker, Dirk.   II. Series.
301—dc21   99-21522

∞ This book is printed on acid-free, archival-quality paper.

Original printing 1999
Last figure below indicates year of this printing.
08 07 06 05 04 03 02 01 00 99

Designed by Janet Wood. Typeset by Robert C. Ehle in 10/14 Sabon.

The authorized representative in the EU for product safety and compliance is:
Mare Nostrum Group
B.V Doelen 72
4831 GR Breda
The Netherlands

www.ingramcontent.com/pod-product-compliance
Lightning Source LLC
Chambersburg PA
CBHW030340240426
43661CB00052B/1694